INTERACTION MATTERS

FROM MATERIALS TO MIND

GOUTAM GHOSH, Ph.D.

First published in 2021 by
BecomeShakespeare.com

One Point Six Technologies Pvt Ltd,
123, Building J2, Shram Seva Premises,
Wadala Truck Terminus,
Wadala (E), Mumbai - 400037
T:+91 8080226699

All Illustrations by the author himself. The cover page concept (front and back) by the author and designed by Aditi Shah

ISBN: 978-93-90543-83-0

CONTENTS

DEDICATION

I dedicate this book for
my father & father-in-law,
whom I lost during this journey

ACKNOWLEDGMENT

I wish to thank my dear readers for their absolute patience in reading and analysing every thought of mine written in this book.

My regards to Ms. Mital Rodrigues, Publishing Associates of BecomeShakespeare.com, for her immediate action after my email to them regarding my book publication. She talked with me friendly. Then I received the call from Senior Publishing Associate, Mr. Rohan Patil. He listened to me with patience and sent publishing details by email. Afterwards my interaction with him became friendly. I got all the necessary support from him. I am thankful to him. I am grateful to Ms. Miral Bheda, my project manager, for cordial interaction and all necessary helps and suggestions. I am also grateful to all members of becomeshakespeare.com for their kind supports.

I am thankful to my colleague, Dr. S. K. Deshpande, for several illuminating discussions, which helped me clarify a few doubts in my mind. I would also like to thank my other colleagues with whom I participate in various debates regularly that enrich my knowledge.

I would also thank Geetanjali and Sanchali, my niece, and daughter, respectively, for fruitful discussions on a few topics related to biology.

I sincerely acknowledge my late father, who taught me rational thinking. I avow my mother, who was always supportive of me. I gratefully admit my brother's contribution in my life as a

guide, my sister, who supports me like a friend. I also acknowledge my late father-in-law, who was a teacher and taught me patience. I sincerely appreciate my mother-in-law, as well as my sister-in-law, for their motherly care when I visit them. I also thank my other near and dear ones who interact with me from time to time encourage.

Last but not least, I gratefully acknowledge my wife, Falguni, for her continuous support and encouragement to my academic activities and personal hobbies.

THE BEGINNING…

WHEN THE UNIVERSE WAS NOT THERE, MATTER WAS ALSO ABSENT. Who created *matter*?

GOD knows!

During this journey from the BIRTH OF THE UNIVERSE until today, our experience through various phenomena would uncover the answer.

We experience different types of interactions through everyday activities such as playing music, reading novels, watching movies, and talking with others. Those interactions are fascinating to me. One day an old friend of mine asked me, *'Why don't you write a book?'* It was so sudden, that I could not answer first! I only asked him back, *'What?'* He told me, 'Yes, you are a scientist and working on interactions in matter. *Don't you have anything to share?'* This conversation opened my mind. I decided to write this book on interactions from materials to mind.

The appellation 'interaction' has originated from Latin words *inter* and *ago,* which means *between* and *to act,* respectively.

What is interaction? According to the Oxford dictionary, it is a way in which matter, fields, and atomic and subatomic particles affect one another. According to the Cambridge dictionary, it is an occasion when two or more people or things communicate with or react to each other. Science claims, *all interactions in this universe are the compositions of four fundamental forces: strong, weak, electromagnetic, and gravitational forces.*

'When I was a graduate student, a' friend gifted me the best-seller book of Professor Stephen Hawking, *A brief history of Time: From Big Bang to Black Holes.* While reading it, my attention got glued to every word on every page. That is the power of a gratifying presentation that can interact with the readers' minds. Professor Hawking has explained the four fundamental forces in his book with a very lucid language that helped me understanding those fundamental forces clearly.

Are there any other kinds of interaction besides fundamental ones? For example, *FRIENDSHIP, PARENT-CHILD,* and *TEACHER-STUDENT* relationships. What kind of force does drive these relational interactions? We shall investigate all these questions at the end of this journey.

The universe got created through a massive explosion, which has named *BIG BANG* by the British astronomer Fred Hoyle (Burbidge, 2003). *THAT EVENT,* around 13.6 billion years ago, had initiated the stories of galaxies, stars, planets, matter, forces, and interactions!

There had been several *EVOLUTIONS* in the universe before reaching its present grand form. A tentative timeline of those evolutions, in Figure 1, depicts rapid *INFLATION* of the space between 10^{-43}–10^{-36} seconds; matter formed after 10^{-32} seconds.

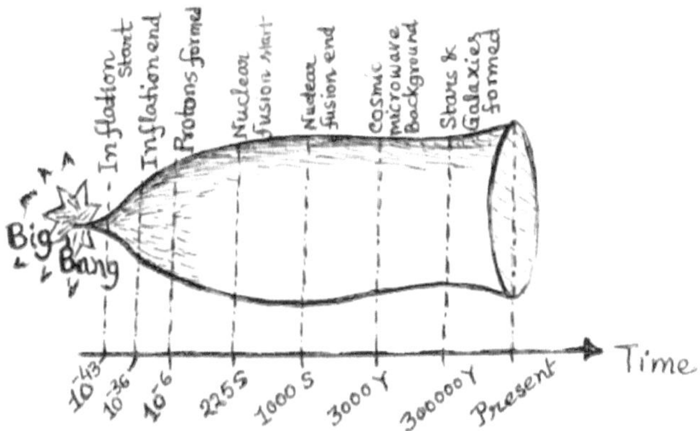

Figure 1. Timeline of the evolution of the universe.

What caused the big bang and formation of the universe? There are many debates on that! Many people believe that God did everything, while others think the coexistence of all favourable conditions caused the sustenance of the universe after the Big Bang. Such massive explosions do occur regularly in the cosmological arena, but a multiverse could not form due to the absence of all favourable conditions in other explosions.

As per the Big Bang theory, there was a highly condensed tiny particle that exploded and created the universe. Based on the current scientific knowledge, we can imagine that a mighty attractive force was densifying that particle. Under an extremely high-pressure situation, several billions of atoms within that particle exploded and released tons of energy. We are assuming that atoms were existing in the pre-universe era. The attractive force was holding the integrity of that particle. If that integrating force was God, it failed eventually! Einstein exclaimed,

"What really interests me is whether God had any choice in the creation of the world".

According to a second version, two giant black holes moving in their trajectories, collided and smashed into small pieces that have composed the universe. However, this theory has not been accepted as popularly as the Big Bang theory.

Though Einstein involved God, Stephen Hawking and Lawrence Krauss claimed in their books, *The Grand Design* (2010) and *A Universe from Nothing* (2012), respectively, that *God was redundant in the creation of the cosmos.* Let us believe that God exists. In that case, THREE questions one may ask:

1. What was the attractive force in the pre-universe particle?

2. Does that force exist now?

3. What is the identity of God?

We will be investigating these questions through this journey.

The Big Bang energy created electromagnetic waves, as well as elementary particles like electrons, quarks, and gluons. Those elementary particles created atoms, and atoms formed all materials that exist in this universe.

The Big Bang has also generated several shock waves in space. More than a century ago, while preparing his general theory of relativity, Einstein predicted that those shock waves collided with each other and created a *gravitational wave* throughout the universe. His prediction has been proved correct by a group of scientists led by Rainer Weiss, Kip S. Thorne, and Barry C. Barish. They detected the gravitational wave using the high-power LIGO (Laser Interferometer Gravitational-wave Observatory) detector, which awarded them the Nobel Prize in 2017.

According to Newton's universal laws of gravitation (published in 1687), two masses always attract each other, but why? Newton did not ask himself the same. But, long after that discovery, Einstein answered that question! He explained that *gravitational attraction originates from gravitational wave and space-time curvature.* His concept has rooted in the quantum mechanical concept.

Space is like a rubber sheet. In the presence of a massive body, it gets curved down and causes surrounding lighter objects to fall into it.

We have learned in the undergraduate course that a force is a spatial gradient of a field. Hence, the gravitational force is a derivative of the gravitational field. The force field has a manifestation of waves, which can interact with each other.

Can the fundamental forces be unified into one form? Scientists do believe that those fundamental forces are the constituents of a unified force. And, assumedly, that unified force was the same that existed in the tiny particle before the Big Bang.

At the Big Bang, that unified force lost its manifestation due to the lack of spatial dimensionality but reappeared after the Planck epoch. Thus, we can be almost sure that the physical forces and physical laws existed even before the creation of the universe.

The inflation of the universe started at 10^{-43} seconds when all physical laws came in action again. Inflation is the rapid expansion of the universe. That inflation started within a billionth of a billionth (that defined 10^{-43} second) fraction of a second after the Big Bang. The time till 10^{-43} seconds is known as the *Planck time,* and the era is the *Planck epoch.* During that time, the quantum gravitational force was so strong that other fundamental forces got amalgamated with it in a unified form. By extrapolating the same toward the pre-universe era, we get the identity of the attractive force in the tiny particle.

After the Planck epoch, the universe got cooled down rapidly due to inflation, and the unified force broke into separate fundamental forces through a process known as the *electroweak symmetry breaking.*

Scientists have now realised that physical laws are meaningless without involving consciousness. All materials are associated with universal consciousness.

What is consciousness? The great Indian poet, Rabindranath Tagore wrote in his poem, *Ami* (*Me*, translated by Rajat Dasgupta in *The Eclipsed Sun*),

> With my senses' hues
> Emerald as green I muse
> And the coral as red;
> As my sight I spread

> The sky is luminous
> East to West with light glorious;
> To rise I said, "Bonny is thee"
> And so did she be!

According to Gautam Buddha,

"Matter is recognised by consciousness only; i.e., there is nothing if there is no consciousness."

Famous scientist, Professor John A. Wheeler said,

"Every piece of matter contains a bit of consciousness".

Michael Mamas wrote in an article entitled, *The Big Bang Theory—A New Perspective,*

"It is fascinating to note that a number of the leading modern physicists whose focus is this Big Bang theory speculate that the unified field is, in fact, nothing more than consciousness."

J. S. Keen wrote that like the structure of the universe, and the laws of physics, consciousness was also created at the time of Big Bang, which relates to the structure of the universe and the physical laws (Keen, 2013). Tagore further wrote,

"…when the creation was new and all the stars shone in their first splendor, the gods held their assembly in the sky and sang, 'Oh, the picture of perfection! The joy unalloyed!'"

Why Gods needed so long time to create the picture of perfection in the sky and on Earth? You find the answer!

The universe contains galaxies and voids. There are objects

in the size range between few hundreds of light-years[1] down to a few millionths of a meter. Novelist, Peter de Vries wrote,

"The universe is like a safe to which there is a combination, but the combination is locked up in the safe."

Matter first started forming as elementary particles like electrons, quarks, and gluons. Those elementary particles are the fundamental constituents of all nonliving and living objects. Around 3.5 million years ago, there were only unicellular organisms on Earth. Through complex interactions, those unicellular organisms transformed into multicellular organisms. The evolution of life is an extraordinary phenomenon, as Pierre Durand, a researcher in the Department of Molecular Medicine and Hematology, The Evolutionary Studies Institute at Wits University, said,

"The evolution from unicellular to multicellular life was a big deal. It changed the way the planet would be forever. From worms to insects, the dinosaurs, grasses, flowering plants, hadedas and humans, you just have to look around and see the extraordinary forms of multicellular existence".

The environmental change was a prime factor for the evolution of life and the enormous biodiversity. There were adverse environments that caused the extinction of some species like dinosaurs. Thus, environs do matter in our lives, as our organs continuously interact with nature.

All interactions come under two broad types: materialistic and nonmaterialistic. Materialistic interactions, such as chemical, physical, or biological interactions, are driven by the fundamental forces and bound by classical space-time dimensions.

Nonmaterialistic interactions such as personal relationships, on the other hand, are not held by such boundaries. Do the fundamental forces drive those interactions? Both interactions connect lives with nature to promote evolutions. Joseph Campbell, an American professor of literature at Sarah Lawrence College wrote,

"The goal of life is to make your heartbeat match the beat of the universe, to match your nature with Nature."

Earth is continuously evolving since her birth. John Muir, a Scottish-American naturalist, wrote,

"This grand show is eternal. It is always sunrise somewhere; the dew is never all dried at once; a shower is forever falling; vapor is ever rising. Eternal sunrise, eternal sunset, eternal dawn and gloaming, on sea and continents and islands, each in its turn, as the round earth rolls."

Weather also changes cyclically. The American writer Mark Twain wrote,

"In the spring, I have counted 136 different kinds of weather inside of 24 hours."

In the universe, the elementary particles formed first, and subsequently, those elementary particles interacted together to form atoms; atoms form molecules. The molecules of life did form only when Earth got the right kind of environment.

The cycles of evolution created *Homo sapiens*, wise man, who has the most complicated physiological systems and brain. How do the human body response to medicines and the environment, and their brain function?

Recently, a new submicron technology, called nanotechnology, has emerged, which has gifted us with many sophisticated gadgets that have improved our lifestyles. Therefore, nanoscience plays a significant role in basic and applied researches.

The concept of energy-matter duality helps us to cross the boundary of the classical domain to the quantum domain and understand various non-materialistic interactions. For example, why do we enjoy music? The mind-brain coordination has a far-reaching role in those interactions.

The questions posed above got explored in subsequent chapters to reach a conclusion about the identity of God.

CHAPTER ONE

FORMATION OF MATTER

PURPOSE OF THIS CHAPTER: To know how do fundamental forces work in building up the universe.

After Big Bang: Age of leptons

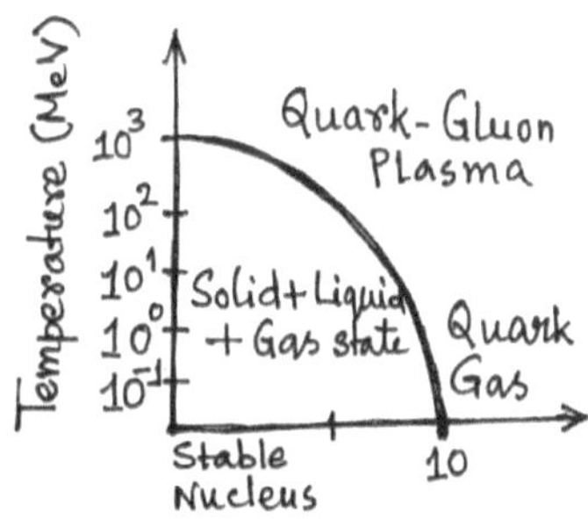

Figure 1.1. Conjectured phase diagram for the state of matter, presented with modification as given in the book of Siemens and Jenson, 1994.

How did matter form and evolve out of nothing? After the Big Bang enormous energy was released; so, the universe was very hot but cooling down very fast due to the inflation. After about 10^{-32} seconds, the spread-out fire of energy started transforming into matter in the form of elementary particles such as *electrons*, *quarks*, and *gluons*. Those are very tiny particles and are called quantum particles, as they never have any well-defined position or momentum.

The transformation of energy into matter gets realised through energy-mass equivalence relationship: $E = mc^2$, where m is the rest mass of an elementary particle, and c is the velocity of light. At low temperatures, free elementary particles are non-existent. But at the very high temperature of the universe, those newly created elementary particles freely existed; such a state of matter is called *quark-gluon plasma* state. Scientists can now generate plasma in laboratories by ionizing gas in a closed chamber, but not quarks or gluons. Creating quarks and gluons from the matter will be a billion-dollar experiment, which had occurred naturally in the baby universe without any scientist! That was the beginning of the formation of matter.

We learned about quarks and gluons very recently. The name *quark* was first coined by the Caltech physicist Murray Gell-Mann, who won the Nobel Prize in physics in 1969 for his work on fundamental particles. A book by Siemens and Jenson (1994) has a conjectured phase diagram of the matter; a rough representation of that has been drawn and shown in Figure 1.1. The elementary particles are the basic units of matter.

PHOTON TO ELECTRON TRANSFORMATION: While traveling at high speed, through a strong electric field and extremely high temperatures (e.g., $7 - 10$ trillion °C $\approx 10^{13}$ °C), photons radiate energy and transform into low energy γ-rays. Those γ-rays decay into matter-antimatter pairs, as $\gamma \rightarrow e^- + e^+$, where e^- represents an *electron*, i.e., a matter, and e^+ represents a *positron*, antimatter. It is known as *pair production*. As the rest energy (energy at rest) of e^- or e^+ is 0.511 MeV, the minimum energy required for a photon to decay into e^--e^+ pair is 1.022 MeV. The strong electric field in the early universe got produced by the light itself, which was a part of the energy released in the Big Bang. Thus, energy had converted into matter spontaneously. Igor Kostyukov of Institute of Applied Physics of the Russian Academy of Sciences (IAPRAS) said,

"A strong electric field can, generally speaking, 'boil the vacuum,' which is full of 'virtual particles', such as electron-positron pairs. The field can convert those types of particles from a virtual state, in which the particles are not directly observable, to a real one."

Age of nucleons

Electrons and positrons are *leptons*; the name came from the Greek word *leptos*, which means small. Quarks and leptons also follow the *Fermi-Dirac statistics* of the spin (s) ½ particles and, thus, belong to another broader family, called *fermions*. They never participate in strong interactions. On the other hand, gluons are *bosons*, as they follow the *Bose-Einstein statistics* of the spin-1 particles; they are much heavier than leptons. The term 'spin' does not refer to a physical rotation, like a spinning top, but a quantum state of matter.

The time between 10^{-35}–10^{-6} s was the *age of leptons* (Figure 1), as leptons such as the electron, muon, tau, and their antiparticles, and neutrinos did form then, in the quark-gluon-plasma soup. Muon and tau are heftier particles having the charge same as $-1e$; neutrinos are neutral particles. Depending on their intrinsic properties (i.e. charge, spin, and mass), quarks have six *flavours: up, charm, top* (each of charge $^{+2e}/_3$, but increasing mass) and *down, strange, bottom* (each of charge $^{-1e}/_3$, but increasingly massive) (Nave, 2008). 'Flavour refers to a particular quality of character' (Cambridge Dictionary). Further, each flavour has three colours: *red, green,* and *blue,* although those are not real colours but the non-Greek nomenclatures. During 10^{-6}–1 s, the universe became further cool motivating quarks (and antiquarks) to interact among themselves and compose bigger particle (or antiparticles). It is called *the confinement of quarks,* which is mediated by gluons. As the mediating particles (e.g., gluon) are undetectable in any interaction, they are called *virtual particles.*

Figure 1.2. Cartoons show quark compositions of (a) proton and (b) neutron

NEUTRON AND PROTON: The cartoons, in Figure 1.2, show the confinements of quarks; two *up* quarks (*u*) and one *down* quark (*d*) form a proton of charge +1*e*, and two *down* quarks and one *up* quark form a neutron of charge zero. Each confinement contains quarks of three primary colours (red, green, and blue) to provide the manifestation of a colourless or white particle.

Confined quarks can change their flavour; thus, neutrons can transform into protons and vice-versa. Such transformations occur in nuclear decays of the radioactive elements. Most of the other particles that did form through confinements of quarks disappeared almost immediately due to a fugitive life, e.g., of the order of 10^{-11} s or less.

NUCLEUS: When the universe became further cool, protons and neutrons interacted and form the nucleus of size about 10^{-15} m. Thus, they are known as *nucleons*. The age of nucleons had a span of 10^{-6}–225 s. An antiparticle is composed of three antiquarks. The antiparticle of the proton is called antiproton.

The confinement of quarks is a short-ranged strong interaction as it gets mediated by the strong force holder gluons. But the strong force becomes weak at very high temperatures. Hence, quarks could exist freely in the extremely hot primitive universe. Within the confinement, the force between the quarks remains unchanged when they move away from each other, but the energy of the mediating field of gluons increases that restrains quarks get apart.

The study of strong interactions in quark confinement is known as the *quantum chromodynamics* (QCD). This subject has partly developed by MIT Professors Jerome Friedman and Frank Wilczek, which brought them the Nobel Prizes in Physics. The Greek word *chrôma* means colour. As quarks have labels of three colours, the term chromodynamics appears in the study of quark confinement. Quarks and gluons are the elementary particles of matter but undetectable; thus, atoms are the detectable building blocks of matter.

ENERGY–MASS–ENERGY CYCLE: With the expansion, the universe further cooled down. Billions of particles and antiparticles lost their thermal motions, came closer, and collided. Consequently, those particles and antiparticles got annihilated, producing colossal radiation. For example, electron and positron annihilate upon collision and form a photon of energy 1.022 MeV. Few annihilations do produce some massive particles of quarks such as B mesons. So, there was an *energy-mass-energy cycle* that took place until about 225 seconds after the Big Bang. Such energy-mass-energy cyclic processes also happen in our ecosystem. For example, plants absorb sunlight and produce food. It is an energy-to-mass transformation. When animals eat vegetables, they generate energy in their body. It corresponds to a mass-to-energy conversion. The burning of coal or petroleum is another example of mass-to-energy reformation. After those annihilations, a small number of particles (i.e., quarks and electrons) had left to form all materials in today's universe.

HOW DID MATTER REMAIN BUT NO ANTIMATTER? As there was an equal number of particles and antiparticles, in the beginning, they should have disappeared due to one-to-one collision. BUT NO, *matter was still present*! It was possible due to

the *matter-antimatter asymmetry*. Professor Stephen Hawkins has explained it in his book, *A brief history of Time: From Big Bang to Black Holes*.

Briefly, when the temperature was very high, more number of positrons (antiparticles of electrons) got converted into quarks than the number of electrons turned into antiquarks. Thus, after extinction, an excess of quarks and electrons had left. However, a large percentage (about eighty-five percent) of those materials is missing there, which has a designation as *DARK MATTER*. Those got constituted of weakly interacting massive particles (WIMPs), a few hundred times heavier than proton or neutron. As dark matter does not interact with light or any other electromagnetic radiation, they remain undetected with existing scientific instruments.

Age of nucleosynthesis

How did big objects form? The answer lies in the cosmologists' model, called *hierarchical structure formation*, which assumes that sizeable bodies got formed from small particles. As mentioned earlier, quarks and gluons interacted to form bigger particles like protons and neutrons, which are known as the *baryons*. Due to spin- property like an electron, baryons are fermions. Baryons do also belong to the family of *hadrons*, i.e., strongly interacting particles. Protons, neutrons, and electrons together form atom.

The term 'atom' has originated from the Greek word *atomos* that means *undividable*. But atoms are dividable; thus, it got the wrong label. However, it happened due to a lack of knowledge.

In the *hierarchical* process, a nucleus form first. Accordingly, few nuclei (plural of nucleus) of light atoms formed within 3–20 minutes of the birth of the universe. This period is called the *age of nucleosynthesis* (see Figure 1). The age that came first after Big Bang is called *Big Bang nucleosynthesis*. Nuclei of heavier atoms formed from hydrogen and helium nuclei in a period called *stellar nucleosynthesis*.

In the age of nucleosynthesis, electrons were freely flying particles due to the very high temperature of the universe with velocity, $v = \sqrt{2k_B T/m}$. Thus, the light particles (m ≈ 0) like electron acquired very high random speeds leading to their free existence. Newly formed protons, neutrons, and nuclei were also existed free at that time, but having slower thermal speeds due to a heftier (~1836 times) mass than an electron.

As the universe cooled further, negatively charged electrons lost thermal motion, clustered either around the positively charged protons or around the positively charged nuclei due to the *electrostatic attraction*, and started revolving. Those clusters were the first atoms, i.e., the embryos or the building blocks of matter in the universe.

Afterward, the gaseous composition of about seventy-five percent of hydrogen atoms, twenty-four percent of helium atoms, and a few others within one percent did form. With the progress of time, those gaseous contents made several clusters of stars, planets, and galaxies that shaped the universe into '*the picture of perfection!*'

Formation of atom

The first atoms formed in the universe were hydrogen (H) and helium (He), the lightest elements. *How do they look?* See the cartoon below.

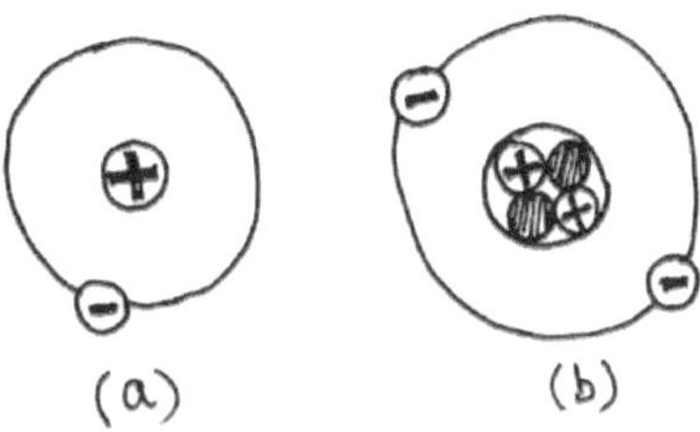

Figure 1.3. Schematic diagram of (a) hydrogen atom (H_1^1), and (b) helium atom (He_2^4)

PICTURE OF ATOM: An atom consists of a positively charged nucleus, the epicentre of power, with few electrons revolving around it, as shown in Figure 1.3. The electronic path of revolution, called the *stationary orbit,* has well-defined energy. As a nucleus is composed of protons and neutrons, it is a massive enclosure, the bulk of the atomic mass (more than ninety-nine percent) has pinned into that minuscule centre. This anatomy, as proposed by the Danish physicist Niels Henrik David Bohr, is known as the *Bohr's atomic model.* He received the Nobel Prize in Physics in 1922 for this work.

SYMBOL OF ATOM: Atom is a neutral entity, so there are as many electrons as the number of protons in the nucleus. This number is called the *atomic number* and denoted by the letter Z. The sum of the proton number (Z) and the neutron number (N)

in a nucleus is known as the *mass number*. It is denoted by the letter A, i.e., $A = Z + N$. An atom (say, X) gets symbolized as X_Z^A, where A and Z both vary from one to the other atom as seen in the Periodic Table of elements.

HYDROGEN, ITS ISOTOPES AND HELIUM: The first atom in the periodic table, hydrogen (H), has only one proton but no neutron in its nucleus ($A = 1$) and one electron ($Z = 1$), see Figure 1.3. Hence, it gets the symbol (H_1^1). If two or more atoms have the same atomic number but different mass numbers, they are called *isotopes*. For example, a hydrogen atom (H_1^1) has two isotopes, deuterium (H_1^2) and tritium (H_1^3). The nuclei of deuterium and tritium have one and two neutrons, respectively, more as compared to that of hydrogen. But tritium is not as stable as deuterium, and its lifetime is very short. Therefore, rarely present in nature.

The second atom in the periodic table, helium (He), has two protons and two neutrons in its nucleus ($A = 4$) and two electrons ($Z = 2$) outside. Thus, represented as (He_2^4).

There is a limit in electron number to an orbit. According to Bohr's atomic model, a maximum of $2n^2$ electrons can get rooms in each orbit (also called a *shell*), where n is the orbit number. For the first orbit, called the K shell, $n = 1$, which can have a maximum of two electrons. The second orbit, called the L shell with $n = 2$, can have a maximum of eight electrons, and so forth.

Usually, an atom has an equal number of protons and neutrons in its nucleus, which is the same as the number of electrons. The nuclear diameter is approximately 10^{-15} meters,

whereas the atomic diameter is around 10^{-10} meters. That is, the nucleus is roughly 10^5 times smaller than the atom where most of the mass has accumulated. Or, in other words, there is a lot of emptiness inside the atom. If you imagine a huge empty football stadium as an atom, then the nucleus would be an iron marble at the centre of the stadium.

HISTORY OF THE FIRST ATOMS: Around 380,000 years after the Big Bang, when the temperature of the universe dropped to nearly 3000 °C, electrons' thermal speed slowed down. As positively charged protons or nuclei were also around, electrons got attracted to those nearly two-thousand times massive particles due to the electrostatic attraction. To avoid collision started orbiting around them. Such combinations formed atoms. In the beginning, atoms of hydrogen, helium (in 3:1 ratio), and a trace amount of lithium got formed.

AND, THEN MATTER: Slowly, all materials in the universe got formed out of atoms. How can planets, stars, or even microorganisms be made up of those tiny particles? Just think about our hair, which is only about 10^{-4} m, i.e., one-tenth of a millimetre thick, but 10^6 times bigger than an atom. Yet all materials, irrespective of their size and shape, are made up of atoms only. *GUESS HOW!*

FORCE WITHIN AN ATOM: We shall see that the electrostatic force, which is also called the *Coulomb force*, plays a fundamental role in forming atoms. Its mathematical form is:

$$F(r) = -\frac{ZQ^2}{r^2} \tag{1.1}$$

Q represents the value of charges. In Eq. (1.1) the negative sign implies the force is attractive, i.e., between a negative charge $-Q$ and a positive charge $+Q$. In the case of an atom, $-Q$ represents the total charge of electrons and $+Q$ the accumulated charges of protons in the nucleus. Z represents the atomic number, and r is the radius of the atom.

The Coulomb force between the charges establishes invisible *lines of coercion,* as shown in Figure 1.4. The number of lines per unit area indicates the strength of the electric field. Conventionally, lines of force emerge from the positive charge and merge into its negative counterpart, as though a give-and-take policy of close attachment. Since it is a bound state, the net energy of the system is negative. It is the system's *potential energy.* If there are 'like' charges, i.e., both are either positive or negative charges, the system becomes repulsive, and the net energy becomes positive. That situation prohibits forming an atom.

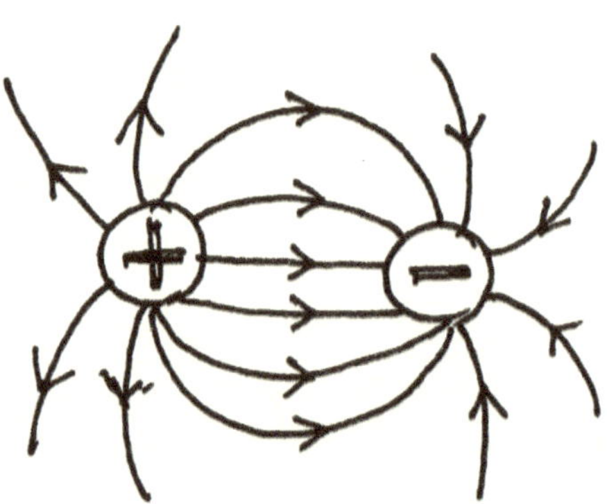

Figure 1.4. Schematic representation of electrostatic lines of force between a positive and a negative charge

The figure also suggests that an atom can get polarised in a strong electric field. Then it becomes a *dipole* with the dipole moment $P = Qr$ where Q (= Ze, e is the charge of each proton or electron) represents the charge at each side of the dipole. The nucleus creates the positive side, and all electrons the other end with a separation of r is the distance between the poles. Such an atom can generate an electric field, which an unpolarised one cannot do. Different atoms have different polarisability due to their size difference that gives materials diverse electrical properties. The electrostatic force creates such diversity.

The interaction between any two subatomic particles gets mediated by an exchange particle, called a mediator. This concept was introduced in 1948 by the American physicist Richard Feynman and is known as *Feynman diagram*. It explains why two particles in an atom do not collide with each other and annihilate. However, collisions and annihilations do occur between particles and antiparticles. For example, the electron-positron annihilations. Thanks to the *matter-antimatter asymmetry* in the universe by the grace of God! Otherwise, we would not have existed today.

BILLIARD BALL FALLACY: Here the question is, can macroscopic bodies vanish on collision? Imagine a billiard game, a ball hits another, and both disappear! But it does not happen anyway though billiard balls also contain subatomic particles, i.e., electrons and protons. So, what averts them from extermination? Think for the answer now, or else we will come to know it through our journey.

MORE ABOUT COLLISION: There is one more point noticeable in the billiard ball collision. Since it is nearly elastic, there is no or minimal loss of energy. So, according to Newton's laws of motion, the *total momentum (mass multiplied by the speed of the body) of the system is conserved*. That means the combined propulsion of the two balls does not change by the collision. With that prefix, it is possible to determine the direction of the recoil paths of the two balls. The same is true for the Feynman diagram.

SYMMETRY IN NATURE: Here, especially note that the physical laws such as the conservation of momentum are consequences of symmetry in the system. If, for any reason, that symmetry breaks, then those physical laws will no longer be valid within that system. Einstein considered the symmetry principle as the primary feature of nature that restricts acceptable physical laws.

The Feynman diagram of the electrostatic interaction between a proton and an electron in the hydrogen atom has shown in Figure 1.5, where a virtual photon (the dashed line), called γ-*particle*, is the mediator. Electron gives away the photon[2] and recoils, while proton instantly absorbs it from a distance and recoils. The recoiled paths get directed by the conservation of the total momentum of the system.

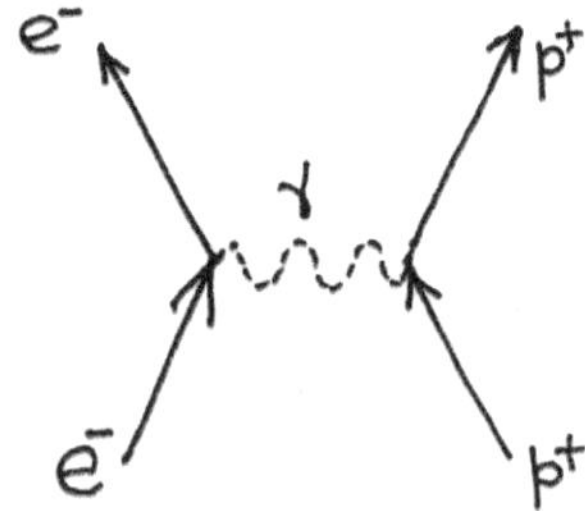

Figure 1.5. Feynman diagram for proton-electron electrostatic interaction in the hydrogen atom

Though the electrostatic force decays inversely with the square of the distance (r) as given in Eq. (1.1), its effect may reach up to infinity. Thus, one charged body can influence another one at an infinite distance away.

The above event appears unfeasible. However, that becomes possible if the virtual photon has no load so that it can travel at a cosmic speed ($v = \sqrt{2\times K.E./m}$) even with infinitesimal kinetic energy. Thus, the electrostatic force creates long-range interactions.

YOU CAN DEMONSTRATE ELECTROSTATIC FORCE. Yes, you can! Take a comb and rub it with your clean hair, then bring it to a piece of paper that will get attached to the comb. So, there is an electrostatic attraction. What is its origin? Again, it arises because all materials are made of atoms. While rubbing, the comb gets electrified due to the accumulation of electrons at its surface. Then it polarises the piece of paper, on touch, with positive charges near to the comb and negative charges rear. That fabricates an electrostatic attraction between the two bodies.

Atoms, in a macroscopic solid, are tightly tethered by physical bonds. Electrons are tied with nuclei in the atoms by the electrostatic bond. Such a multiple bonded cohesive system has immense potential energy, and the surface of the body has a potential barrier.

ELECTRON-FLOW FROM THE RED-HOT IRON SURFACE: You may have seen a wave-like flow from the surface of a red-hot iron (made of iron atoms). It is a stream of electrons ejecting out of the surface atoms of the metal. *How do electrons surmount the potential barrier at the metal surface?* It is because the outermost electrons of the surface atoms absorb positive energy on heating and overcome the attractive potential energy within the atoms causing them to leave the metal surface.

Why do ejected electrons appear as a wave? That is because *quantum particles, like electrons, can also manifest as waves.* The associated wavelength (λ) is, according to the de Broglie (Louis de Broglie, a French physicist) equation, $\lambda = h/p$. In this equation, p is its linear momentum, and h, the Planck constant. It is known as the *wave-particle duality* that has introduced a new form to the electronic orbits.

ELECTRONIC ORBIT GETS A NEW SHAPE: According to the second postulate of Bohr's atomic theory, the angular momentum of an electron in a stationary orbit is equal to $nh/2\pi$, where n denotes the orbit number. Suppose an electron of mass m is revolving at a speed v on an orbital path of radius r. Its angular momentum will be mvr, where mv is the linear momentum (p) of the electron. Thus, $mvr = {}^{nh}\!/_{2\pi}$. using the de Broglie wavelength, we get

$$2\pi r = \frac{nh}{mv} = \frac{nh}{p} = n\lambda \qquad\qquad (1.2)$$

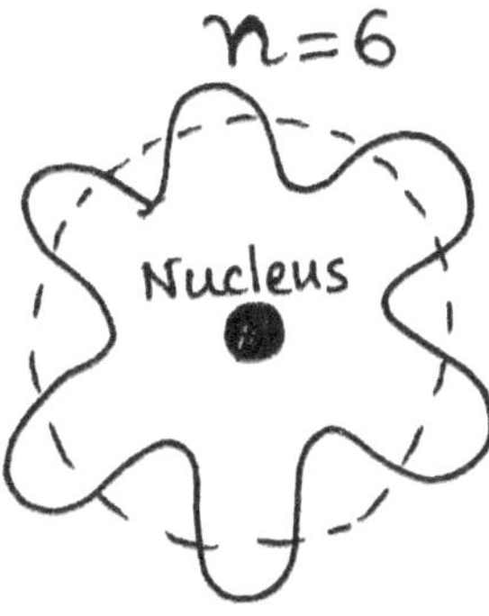

Figure 1.6. Wave representation of Bohr's electron orbit for $n = 6$

Eq. (1.2) implies that the orbit (of circumference $2\pi r$) gets an *undulating shape* of an integer number (n) times the de Broglie wavelength of electrons, as shown in Figure 1.6 for $n = 6$. It is a mathematical interpretation of the electron orbit; the real shape is much different from this. In practice, the shape of an electronic orbit has a flexible form depending on the quantum states of electrons.

The formation of atoms, in the early universe, accompanied the emission of immense γ-rays and cosmic rays. On cooling, cosmic rays did condense into the cosmic microwave background (CMB) radiations, which do exist even today. First, in 1941, Andrew McKellar discovered CMB. In 1964, American physicist Arno Penzias and radio astronomer Robert Woodrow Wilson rediscovered the CMB radiations. They also estimated the temperature of CMB radiations as 3.5 K. This work honored them with the Nobel Prize in 1978.

MORE FORCES WITHIN AN ATOM: As discussed earlier, a helium nucleus has two protons and two neutrons. Protons have positive charges, thus, repelling each other. That should make a nucleus (except hydrogen) unstable! But it is not true; as all nuclei having the neutron-to-proton ratio between 1.1 and 1.5, are stable. How is that possible? It will be answered shortly.

Eq. (1.1) reveals that electrons remain bound within an atom by the electrostatic attraction towards the nucleus. *Why electrons do not jump onto the nucleus, and collapse the atom?* Bohr's model has the answer. Though the nucleus pulls them toward itself, electrons tend to depart away by an opposite but equally strong *centripetal force*. A balance between those two forces forges an atom stable. The centripetal force arises due to the revolution of electrons in stationary orbits as

$$f_c(r) = \frac{mv^2}{r} \tag{1.3}$$

The radius (r) of a Bohr orbit can be obtained by equating Equations (1.1) and (1.3),

Though there are few drawbacks in this simple model, it works well in calculating the energy spectrum of hydrogen gas. However, it is unable to explain the fine-splitting of spectral lines in an electric field (known as the *Stark effect*) or a magnetic field (known as the *Zeeman effect*). It could not also explain the spectra of the heavier atoms. However, the significance of this model lies in its quantum concept of energy, which was novel at that time.

Nucleus

STRONG NUCLEAR FORCE: All atomic nuclei, except radioactive elements, are stable despite proton-proton repulsion. So, there is a super-attractive force that supersedes the Coulomb repulsion. It is the *strong nuclear force*, which is very short-ranged; it works on a length scale of 1.0–2.5fm (1fm, or, femtometer = 1.0×10^{-15} m). The same force becomes repulsive to prevent nucleons from coming closer than 0.7fm and save the nucleus from collapsing. As this force binds all nucleons, it is also known as the nuclear binding force.

How strong is the strong nuclear force? It is so strong that a massive energy gets released when a nucleus split. The splitting of a nucleus is known as *nuclear fission*. It is used for making nuke (that had ruined two cities, Hiroshima and Nagasaki, in Japan during World War II), as well as for producing electricity in nuclear reactors.

SUNLIGHT, THE SOURCE OF ENERGY: If two nuclei are fused, immense energy gets produced. It is known as *nuclear fusion*. But, how could two nuclei fuse despite a strong repulsion between them? It could be possible by applying a massive external pressure, which is arduous by any mechanical process. However, it happens at the core of a star such as our sun. In the interior of a star, a very high density of gases exists that creates a strong gravitational pull. Consequently, the hydrogen atoms bump *en masse*, and their nuclei fuse. In a sequence of such processes, four times heavier helium nuclei get formed, and immense energy produced in the forms of light and heat, i.e., SUNLIGHT.

The plants and animal kingdom on Earth absorb sunlight in the daytime for generating food as well as processing many metabolic activities. So, it is a blessing that the nuclear and the gravitational forces do exist naturally.

NUCLEUS, THE VORTEX OF FORCE: I endorsed the kernel as the epicentre of power (in an atom), where most of the mass gets condensed. It is like the vortex force created in a pool of water, as shown by the cartoon in Figure 1.7. The downward pull at the vortex is quite powerful and localised, creating the epicentre of a high potential gradient, and imitates the strong nuclear force. The radial arrows towards the vortex represent a low potential gradient and a weak force. It mimics the weak electrostatic force between the nucleus and the electrons.

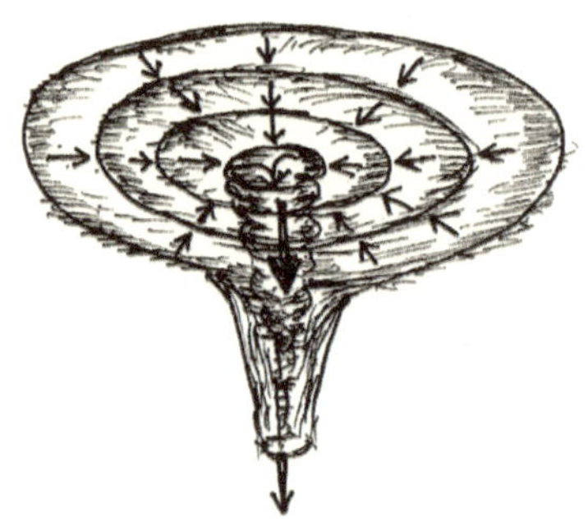

Figure 1.7. Vortex in a water pool

BLACK HOLE: The vortex force in a pond arises due to the Earth's gravitational pull, where there is a pinning of mass. It is also like a black hole. However, a black hole has such a prodigious drag that it swallows even massless photons. As no light can come

out, it appears black.

NUCLEAR BINDING: Where does the nuclear binding force come from? Einstein's mass-energy equivalence has the answer. For example, if you strike two stones, a spark of fire will appear. A loss of minute mass from the rocks gets converted into the flicker. That is what the mass-energy equivalence in far and wide. In a nucleus, masses of nucleons transform into energy. Quantitatively, it is as follows.

Suppose M is the mass of a nucleus, and M_p and M_n are, respectively, those of a proton and a neutron in their free states. Thus, the loss of mass in binding the two nucleons is given by $\Delta m = [(Z \times M_p + N \times M_n) - M]$, where Z is the number of protons, and N is the number of neutrons in the nucleus. Note that in some cores, $N = Z$, while in others, $N > Z$. The quantity Δm is called the *mass defect*.

The mass defect transforms into the binding energy of nucleons according to the mass-energy equivalence relation $E = -\Delta mc^2$. It is the origin of the strong nuclear force, which is charge independent. Thus, the binding force between any two nucleons (e.g., proton-proton, neutron-neutron, or proton-neutron) is the same. So, Einstein's energy-mass equivalence is a powerful equation. It can be used either against or in favor of humanity. We have to decide whether that great discovery should be used as a blessing or as a curse. Einstein himself remarked,

"The release of atom power has changed everything except our way of thinking…the solution to this problem lies in the heart of mankind. If only I had known, I should have become a watchmaker."

What is the exchange particle in the strong nuclear interaction? The German physicist Werner Heisenberg introduced the concept of the exchange force in the nucleus that binds nucleons together. Heisenberg assumed that protons and neutrons are two different quantum states of the same particle[3]. Like the confinement of quarks in proton and neutron by gluon, nucleons get glued within a tiny nucleus by the meson particle called *pion* (π-meson). The Feynman-diagram of this interaction has shown in Figure 1.8.

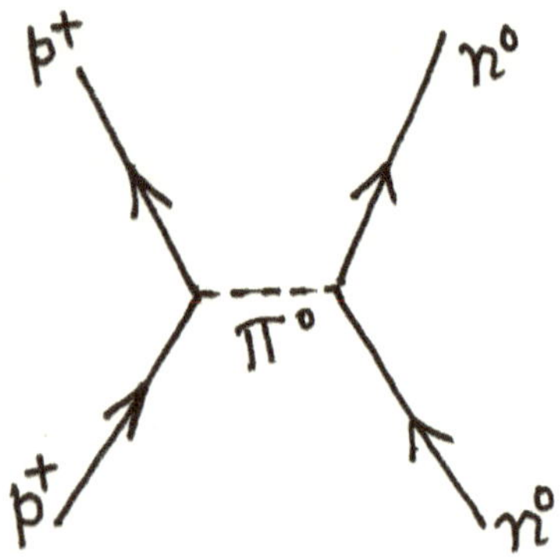

Figure 1.8. Feynman diagram of the proton-neutron strong nuclear interaction

Mesons are composed of quarks and antiquarks with gluons and are massive particles, unlike photons. So, the characteristics of the mediating virtual particles depend on the particles undergoing scattering. If the scattering occurs between the light particles like an electron and a positron, the mediating particle would be a light virtual particle, like a photon. On the other hand, if it is between two massive particles like a proton and a neutron, then

the mediating particle would be a bulky virtual particle, like a meson.

RADIOACTIVE DECAYS AND WEAK NUCLEAR FORCE: Nuclei for which the *N/Z* ratio lies between 1.1 and 1.5 are stable, while the remaining others are unstable. Unstable nuclei decay spontaneously into lighter stable nuclei, with the emission of energy and, in some cases, elementary particles. It is called *nuclear* or *radioactive decay*, which is driven by the *weak nuclear force*. The decaying elements or substances are called *radioactive elements.*

The stability of a nucleus depends on the balance between the two forces, i.e., the electrostatic repulsion and the strong attraction. The equipoise gets disturbed if either neutron or proton number increases beyond their limits defined by the ratio of *N/Z* as that causes either attractive or repulsive force dominating.

There are three types of nuclear decays: *alpha decay, beta decay,* and *gamma decay.* Alpha decay is associated with α-particle emission, beta decay with β-particle emission, and gamma decay with the radiation of γ-particles. Many times, alpha decay has accompanied by the γ-radiation.

Uranium-238 (U_{92}^{238}) is a radioactive element, which decays into thorium-234 (Th_9^{234}) with the emission of an alpha (α) particle as $U_{92}^{238} \rightarrow Th_{90}^{234} + \alpha$. An alpha particle has a mass number four, and an atomic number two. So, it is a helium atom (He_2^4).

A β-particle could be either a beta minus (β^-) or a beta plus (β^+). The (β^-) particle mimics an electron, while the (β^+) particle a positron. A positron is the antiparticle of an electron, i.e., they have an equal and opposite charge but the same mass.

In the decay of a neutron to a proton, a β-particle emits as $n \rightarrow p + e^{-}(\beta^{-})$. In such decay, the atomic number (Z) of the product increases by unity, whereas the mass number (A) remains the same. For example, thorium-234 decays into proactinium-234, as $Th_{90}^{234} \rightarrow Pa_{91}^{234} + \beta^{-}$. But this expression is incomplete. What is missing in this?

As a neutron (M_n = 1.674929×10^{-27} kgs) is slightly bulkier than a proton (M_p = 1.672623×10^{-27} kgs), the nuclear mass of the product element is minutely lesser than that of the parent nucleus. Nevertheless, the mass number (A) remains the same in beta decay. The finite loss in the nuclear mass (= $M_n - M_p$ = 2.3×10^{-30} kgs) is balanced by the emission of a light particle, called *antineutrino* (v), to satisfy the conservation of mass of the system. Thus, the correct expression for thorium decay would be $Th_{90}^{234} \rightarrow Pa_{91}^{234} + \beta^{-} + \bar{v}$. A neutrino (or an antineutrino) is a fermion.

Figure 1.9 shows the Feynman-diagram of beta minus (β^{-}) decay. The directions of the recoiled paths of the emitted particles get fixed by the law of conservation of momentum. The decay process entails the weak nuclear force.

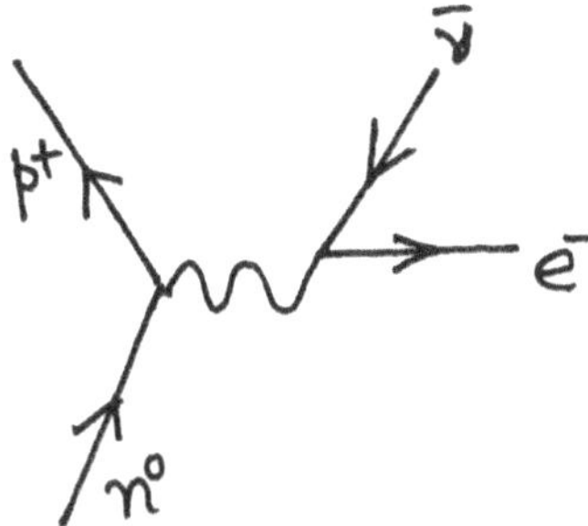

Figure 1.9. Feynman diagram for neutron decay in the nucleus

Beta plus (β^+) decay is another example of the weak nuclear interaction. For example, the transformation of a proton to a neutron. It gets accompanied by the emission of a β^+ particle as well as a neutrino (v) from the nucleus. Sometimes, a proton captures an electron from the innermost K shell (called *K-capture*). Consequently, one of its quarks changes flavour from up (u) to down (d) as $+\frac{2}{3}e\,(u) - e = -\frac{1}{3}e\,(d)$. In this process, the wave function of the absorbed electron collapses to get confined into the proton. The corresponding decay process is $p \to n + e^+(\beta^+) + v$.

Weak nuclear interactions do occur continuously in the core of all stars when hydrogen nuclei transform into helium through the fusion process. The study of weak nuclear interaction is known as the *quantum flavour dynamic* (QFD).

How weak is the weak nuclear force? It is feebler than the electrostatic force, but its interaction range is the shortest among all fundamental forces. It is puzzling! We know that a weak force causes a long-range interaction, whereas a strong one a short-range interaction.

What is the exchange particle for a weak nuclear reaction? The answer to this question was given in 1967 by Professor Abdus Salam of Imperial College of London, and Professor Steven Weinberg of Harvard University in the USA. They proposed that *a weak nuclear interaction gets mediated by multiple exchange particles such as a massless photon and three other massive spin-1 particles* (called *gauge bosons*), *denoted by* W^+, W^- *and* Z^0. Though those hefty particles are weightless at very high energies, become weighty at low energies and make the weak forces very short-ranged. Salam and Weinberg, together with Sheldon Glashow of Harvard University, received the Nobel Prize in Physics in 1979 for this discovery.

How do those gauge bosons[4] acquire mass at low energies? To answer this question, Weinberg and Salam used the symmetry breaking concept, such as the Higgs mechanism of symmetry breaking, in a standard model.

STANDARD MODEL: The standard model reveals that above the critical temperatures, all elementary particles become massless. However, below those temperatures, the Higgs field (a quantum field) causes spontaneous breaking of electroweak symmetry of gauge bosons by condensation resulting in them acquiring masses.

Fermions, such as the leptons and quarks, also gain mass through some interactions with the Higgs field.

EXPERIMENT AT CERN LABORATORY: The existence of those three massive particles (W^+, W^- and Z^0) has been observed experimentally in the CERN laboratory at Geneva in Switzerland. Professor Carlo Rubbia, who led that project, received the Nobel

Prize in 1984 along with one of his team members, an engineer, Simon van der Meer, for that discovery.

FORMULATION OF NUCLEAR FORCES: If a massive particle mediates an interaction, the inverse square law (Eq.1.1) of force becomes inapplicable. Thus, both strong and weak nuclear interactions can't have a formulation similar to Eq. (1.1). Those forces decrease exponentially with the distance ($F\sim e^{-d}$) that is inversely proportional to the mass ($d\sim 1/m$) of the mediating particle. So, the nature of the nuclear forces is different from the other two fundamental forces.

Heavier atoms

The light elements like hydrogen, helium, and a small amount of lithium had formed during the Big Bang nucleosynthesis era; few heavy atoms up to nickel got formed much later through the nuclear fusion of light elements. Most of the other building blocks are created in dying stars through some complex fusion processes. For instance, carbon atoms got synthesised in the red giant stars, as explained below.

With aging, when a star exhausts its storage of hydrogen gas and cools down at its outer region, it appears red. Thus, it becomes a red giant star. The thermal pressure in those stars reduces slowly. Consequently, the core becomes denser as it cools. That creates a gravitational pull towards its centre and initiates fusion between the helium (*He*) nuclei. Through such processes, heavy elements such as carbon do form. The reaction steps are shown below:

$$He_2^4 + He_2^4 \rightarrow Be_4^8$$

$$Be_4^8 + He_2^4 \rightarrow C_6^{12} + \gamma$$

It is known as the *triple-alpha process*, as three helium atoms or alpha particles are involved. The fusion of one more helium with carbon forms an oxygen $_{14}^{28}$ atom. Due to those fusion processes, the core becomes hot, and temperature may reach up to about 10^8 K.

Thus, the constituting elements of our body, such as hydrogen, carbon, nitrogen, and oxygen got formed through various cosmological events, i.e., those are stellar elements.

Age of galaxies and stars

THE PICTURE OF PERFECTION: As time elapsed, gases of atoms and ions, as well as the dust, started clustering due to gravitational attraction. Slowly those clusters formed stars and planets. Several such giant bodies had grouped into supergiant clusters that are known as galaxies. Those developments took place in *the age of galaxies and stars. When the creation was new, and all the stars shone in their first splendor…gods sang, 'Oh, the picture of perfection!'*

The cosmological model that describes the formation and evolution of stars is known as the *Solar Nebula model*. That model was developed in the 18th century by Emanuel Swedenborg, Immanuel Kant, and Pierre-Simon Laplace.

BIRTH OF STAR: The Solar Nebula model says that the

centre of a prototype star keeps growing denser due to gravitational attraction and becomes hotter due to the collision of particles. That process continues until a chain of thermonuclear fusion starts, which produces enormous heat. Consequently, a thermal repulsion grows up, which reciprocates the strapping gravitational pull and stabilises the star.

However, the virial theorem explains that *a twinkling star may collapse if its gravitational potential energy exceeds twice its thermal energy* (Kwok, 2006). Therefore, the criterion of a twinkler to remain stable is that its gravitational potential energy must be less than or equal to twice its thermal energy.

DEATH OF STAR: When a giant star becomes old, exhausts its storage of gases and thermal energy significantly, the star undergoes gravitational collapse and explodes, which is known as a *supernova.*

IT IS THE CYCLE…! A supernova disposes of mammoth materials. Thus, the interstellar medium turns rich with heavier atoms and various other substances that are the ingredients to give birth to new stars. (Krebs and Hillebrandt, 1983). In a new star, the same processes continue. Gravitation keeps on densifying the gas leading to thermonuclear fusions. Hydrogen nuclei fuse to form helium and release thermal energy causing repulsion to counter the gravity. Then the star stabilises.

In the beginning, a weak gravitational attraction delays the formation of a star. Our daystar took about 100 million ($\sim 10^8$) years to grow as a stable stellar body.

These cosmological events keep happening forever with the

assistance of the fundamental forces of interaction. There is no known power to stop those processes.

"A zygote is a gamete's way of producing more gametes. This may be the purpose of the universe" — Robert Heinlein

FORMATION OF PLANET: The formation of planets also has several theories. The most accepted one is known as the *protoplanet hypothesis*. The hypothesis says that tiny objects such as gases and dust particles stick together due to the gravitational attraction and form small planetesimals of diameter about 1.0 km. When those planetesimals collide and stick together, form sizeable planets. Those are protoplanets. The size and mass of a protoplanet are like our moon. Accumulation of more materials onto a protoplanet causes it to grow as a planet. It is called *accretion*. The steps in brief are:

$$Gas + Dust \xrightarrow{Gravity} Planetesimals \xrightarrow{Gravity} Protoplanets \xrightarrow{Accretion} Planets$$

STRUCTURES OF PLANETS: In our solar system, some planets are rocky, and some are gases. Planets (including Earth) that are in closer proximity to the sun are rocky, but those (such as Jupiter) far away are gaseous. As the sun makes a vast region in space very hot, no gaseous cluster could form. However, beyond that region, gaseous planets could sustain. The mammoth orb in our solar system is Jupiter, which is a gaseous composition of hydrogen and helium. It is like a star in its constitution, though not a star. If Jupiter had been about 80 times more massive, it would have become a star rather than a planet due to the effect of strong gravity. Saturn is the second giant planet of gases, mainly of hydrogen and helium.

SO, GRAVITY IS IMPORTANT! *What is gravity?* The gravity or the gravitational force is a *pulling or attractive force* between two masses. Usually, we see a heftier body pulls a lighter one; for example, every object does fall on Earth, the Earth never falls on them. Why?

The answer came out from the discovery of the 'laws of gravitation' by Sir Isaac Newton (1643–1727 AD), and there is a well-liked anecdote behind. In short, when young Newton was sitting under an apple tree, an apple fell in front of him, and in a stroke of brilliance, he got curious, why everything falls? That led to the discovery of gravitation. But the fact is that he was working on planetary motion. After several years of working on mathematical formulations (mainly calculus), he derived the force of gravity. It is inversely proportional to the square of the distance between two bodies and directly proportional to the product of their masses.

Suppose the masses of two bodies are m_1 and m_2, and the distance between them is r, then the force of gravity is given by,

$$F(r) = -G\frac{m_1 m_2}{r^2} \tag{1.4}$$

G is known as the universal gravitational constant, and its value is 6.67×10^{-11} Nm^{2}kg^{-2}. The negative sign refers to the attractive nature of the gravitational force.

You may notice that Eq. (1.1) and Eq. (1.4) have similar forms. However, the electrostatic force contains a product of two charges, and the gravitational force has the multiplication of two masses. Besides, the electrostatic force could be either attractive or repulsive, whereas the gravitational force is always attractive. Like the electrostatic force, the gravitational force also propels long-ranged weak interaction.

Usually, the gravitational force is feebler than the electrostatic force but mightier than others near a massive body like the sun. Like other interactions, gravitational interaction also gets mediated by a virtual particle, which is the spin-2 *graviton*.

STRUCTURE OF GALAXY: A galaxy is a gravitationally bound system of stars, stellar remnants, interstellar gases, dust particles, planets, and dark matter (Sparke and Gallagher, 2000). All those revolve around a giant black hole. Consequently, a galaxy appears like a disk. On average, the diameter of a galaxy is about 200 AU (astronomical unit, 1 AU = average distance between the Earth and the sun). Due to gravitation, a depression forms along the spin-axis at the centre-of-mass of the galaxy. There are a few hundred billion galaxies of different sizes in the universe. *Milky Way* is one of those galaxies in which we live.

Milky Way and Earth

In the night sky, you might have seen a band of light; it is the *Milky Way*. It formed about 4.6 billion years ago, including millions of stars and planets. However, none of them could be distinguished separately with bare eyes. That is why it appears like a white path.

GALILEO'S TELESCOPE: In 1610, the famous Italian physicist Galileo Galilei first observed individual stars in the

Milky Way through a self-made telescope.

The galaxy has a spiral structure due to its continuous spinning around a central point, called the galactic centre. If you take a brush-full of watercolour and start drawing parallel arcs from a central point, it will appear like the Milky Way, as shown in Figure 1.10. It is the same spiral structure as we see in many physical systems. How amazing it looks like, as Joseph Leonard Gordon-Levitt commented,

"The spiral in a snail's shell is the same mathematically as the spiral in the Milky Way galaxy and it's also the same mathematically as the spirals in our DNA. It's the same ratio that you'll find in very basic music that transcends cultures all over the world."

A supermassive black hole, *Sagittarius A-star*, is there at the galactic centre of the Milky Way. There are isolated clouds of dust and gas, formed due to gravity, which are called *nebulae*. Nola Taylor Redd wrote,

"Approximately 4.5 billion years ago, gravity pulled a cloud of dust and gas together to form our solar system."

Figure 1.10. Structure of our Milky Way

OUR SOLAR SYSTEM: The cloud contracted under its own gravity, and our sun formed in the hot dense centre. The remainder of the cloud formed a swirling proto-planetary disk containing planets, moons, asteroids, and other small bodies, all orbiting directly or indirectly around the sun. There are eight planets: Mercury, Venus, Earth, Mars, Jupiter, Saturn, Uranus, and Neptune; and five officially recognised dwarf planets: Ceres, Pluto, Haumea, Makemake, and Eris.

There are billions of similar solar systems in the Milky Way and billions of stars to amuse us. That is why the famous British poet Jane Taylor wrote,

> Twinkle, twinkle, little star,
>
> How I wonder what you are!
>
> Up above the world so high,
>
> Like a diamond in the sky…

How do those stars and planets, including our Earth, hang in space? Earlier, people believed that the Earth was a flat disk on the back of turtles and its turtles till down. Though nobody believes it now, many people think that the sun is stationary and the Earth and other planets are revolving around it, which is not the correct picture. Our solar system, including the sun, is rotating around the master, Sagittarius A-star. The gravitational force acts as a thread to bind every object in the galaxy with the master. *There is unity under gravity.*

The power of fundamental forces is astounding! After the Big Bang, the confining pressure in the pre-universe tiny particle

got invigorated as the four fundamental forces and distributed among all materials throughout the universe.

Fundamental forces do generate many natural phenomena such as thunder and cyclone on stage. However, due to the lack of knowledge, people in the early days thought those natural events as the curse of annoying numen! To get rid of those cursing phenomena, they started praising Him in their ways. Slowly, the image of an unseen superman became God, who was all in all in this universe. Even today, people believe that God is a persuasive person who has control over everything. *It is because of the downloaded information in our subconscious mind from earlier generations* (Lipton and Bhaerman, 2017). What is reality? Let's investigate it through this journey.

"In architecture and interiors, as well as fashion, there is an interaction that is both functional and aesthetic" — Joseph Altuzarra

Evolution of Earth

The Earth formed through the accretion of dust and other debris under the drive of the gravitational attraction. Then she evolved through various cycles to the present form about 4.6 billion years ago, as determined through the isotope dating of rock samples. However, the core had formed much earlier, between 10 and 30 million years after the Big Bang (Cameron 2002), since then advancing continuously. The reason for taking so long time was related to the reduced gravitational force between the Earth's core and other embryonic bodies (Lunine 2006), such as gases and dust particles. Those masses contributed to the remaining part of the Earth. As the Earth grew, its surface, as well as the atmosphere, had to evolve before blooming up life on it. Scientist, environmentalist, and futurist James Lovelock suggested (Lipton and Bhaerman, 2017),

"Earth, itself, is a living entity that uses evolution to regulate its own exceedingly complex metabolism."

EARTH'S STRUCTURE: From a prototype planet, Earth grew layer by layer. There are three main layers from the top of the surface down to the centre: *crust, mantle*, and *core*.

The core has two parts: inner and outer. Though the inner part has solidified due to tremendously high pressure on it, the outer part remained as the liquid. The hard part of diameter about 5300 km, consists of a considerable amount of iron and nickel metals. The outer part of thickness about 2,400 km also contains iron and nickel as major components but in molten form.

The mantle is about 2900 km thick, which mostly consists of oxygen, silicon, and magnesium. It got two layers: the upper layer is solid, and the inner layer consists of molten rocks, called *magma*, as well as some dissolved gases.

The outermost layer of Earth is called the crust, which is 133 km thick and made of solid rocks of lighter elements like silicon, oxygen, aluminum, iron, calcium, and sodium. The crust and the upper part of the mantle are jointly called the *lithosphere*. It consists of large plates, called *tectonic plates*, which float on the magma. The movements of tectonic plates sometimes cause

severe earthquakes and tsunami.

There are two types of crusts, *continental* and *oceanic*. The chemical compositions and physical properties of those two crusts are different as they got formed through independent geological processes. The continental crust, which is about 30–50 km thick, has a granitic layer of less-dense igneous rocks. Those rocks are rich with silicate minerals and molten or semi-molten natural materials like aluminum, sodium, and potassium. The oceanic crust is about 5–10 km thick, composed of magnesium and iron-rich denser rocks.

GEOMAGNETIC FIELD – COSMIC SHIELD: The molten fluid of iron ions at the outer layer of the core has a convection motion due to the spinning of Earth that generates the geomagnetic field. That field can be realised by observing the deviation of a magnetic needle. It creates a magnetic shielding against the cosmic stream of charged particles and protects life on Earth.

People are curious about the inner structure of the Earth. There are many fictional stories on it. The best-known fantasy novel, *Journey to the Centre of the Earth,* is written by Jules Verne. The American director Eric Brevig made a 3D movie based on that novel.

MAGNETIC FORCE AND EXCHANGE PARTICLE: So far, we knew four fundamental forces, namely, *strong, weak, electrostatic,* and *gravitation.* Now, the *magnetic force* has been introduced, which makes a total of five fundamental forces. The electrostatic force controls the interaction between static charges, which is mediated by a virtual photon. When charges (e.g., electrons or ions) are in motion, they produce a magnetic field.

Like the electrostatic force, the magnetic force also has both attractive and repulsive properties. It has the same mathematical form as that of the electrostatic force, Eq. (1.1), except the electric charge gets replaced by the magnetic pole strength. However, unlike an electric charge, no isolated magnetic pole can exist. Always, two opposite poles (north and south) remain in pair as a dipole, which we use as a magnet. Two like poles repel, but opposite poles attract each other.

Suppose, the strengths of two poles are p_1 and p_2, then the magnetic force is written as,

$$F = \pm \frac{\mu}{4\pi} \frac{p_1 p_2}{r^2}$$

(1.5)

The *plus* and *minus* signs in Eq. (1.5) refer to repulsive and attractive interactions, respectively, r denotes the distance between two poles of two nearby magnets, and μ refers to the permeability of the medium. The magnetic interaction is mediated by a spin-1 virtual particle, called *magnon,* which gets exchanged between two spinning electrons in nearby magnets. The concept of a magnon was introduced in 1930 by Felix Bloch (1930).

ATOMIC ORIGIN OF MAGNETISM: As electrons revolve around the nucleus in circular orbits, they possess *orbital angular momentum,* as well as *spin angular momentum* because of their spins. Those motions produce tiny currents and the magnetic field in an atom. Nucleons also have spin motions that create a magnetic field, but that is negligibly small compared to the electron magnetic field.

ELECTROMAGNETIC FORCE: As moving electrons

generate both electric and magnetic fields simultaneously, why should the two be separated? Indeed, those are components of a unified *electromagnetic (e. m.) field,* which gives rise to the *electromagnetic force.*

The electric and the magnetic fields are perpendicular to each other, and both are perpendicular to the direction of propagation of the e. m. wave, as shown in Figure 1.11. The Scottish physicist James Clerk Maxwell published the unification theory of electric and magnetic fields in 1865. He also proved light as an e. m. wave that may have different frequencies and amplitudes. An electromagnetic wave can propagate without any medium. Thus, the number of fundamental forces remained four as:

(i)   Strong nuclear force
(ii)   Weak nuclear force
(iii)   Electromagnetic force, and
(iv)   Gravitational force

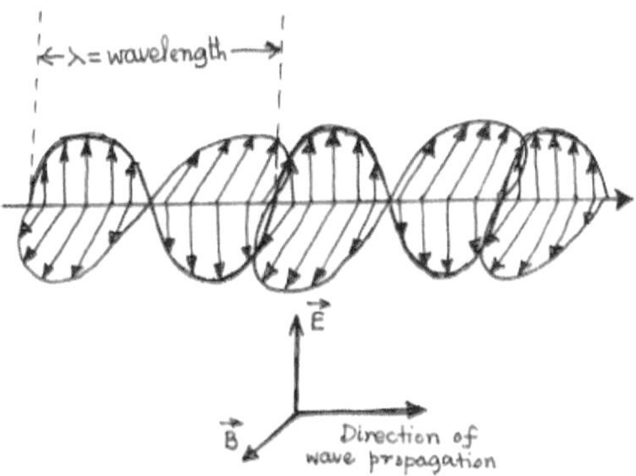

Figure 1.11. Electromagnetic wave propagation showing electric field and magnetic field

In conclusion, after the Big Bang, a large amount of energy was released that partly got converted into tiny particles, such as quarks, gluons, and electrons, and their antiparticles. Bigger particles such as protons and neutrons got formed by the confinements of those elementary particles and antiparticles through strong nuclear interaction. Accumulation of gaseous atoms and ions along with dust could take place due to the gravitational attraction. The fusion between atomic nuclei that produce stars (such as the sun) is possible due to the strong nuclear interaction. Some heavier elements got formed in the stars due to the fusion reactions. The nuclei of some atoms, called radioactive elements, are unstable and decay into stable lighter atomic nuclei due to the weak nuclear interaction. There are electrostatic and magnetic interactions between the charged particles when they are static and in a dynamic state, respectively.

In the following sections, I shall discuss the formation and evolutions of various materials on Earth and the significance of the fundamental forces behind all these.

Evolution of matter on Earth

We know that matter exists in three states: solid, liquid, and gas. But initially, the universe was crammed with ionized gas, the *plasma state* of matter. As the temperature dropped, the ionized state underwent a phase transition and created sequentially diverse forms of matter we see today.

In the beginning, Earth's environment had a gaseous composition of hydrogen and helium in a ratio of 3:1. However, there was no oxygen and carbon dioxide, as well as water to

support life. Furthermore, in the absence of a strong gravitational field, the existing gaseous environment also could not retain for long.

Slowly, as the Earth grew in mass, developed a stronger gravity that helped to capture some gaseous atmosphere of essential gases again. Thanks to the gravitational force. Later on, a boundless ocean of water has also been rewarded, to make her a perfect place for life.

Now, not only a diverse species of life, but there are many beautiful mountains, vast deserts, and sandy beaches, colourful pebbles and crystals, dazzling diamonds, and the strange structure of rocks.

Nonliving matter: Solids

How the Earth crust received so many variations in size, shape, and compositions? Those morphological divergences have occurred in the length scales of few millimetres to centimetres, even up to a few meters to kilometres. Those morphological variations come broadly under two categories: *primary structure* and *secondary structure*.

Primary structures got formed during solidification from melts to either crystalline or sediment morphology. Secondary rock structures got formed from primary rocks due to either compressive or stretching stresses or a combination of both. Depending on the directions of stress application, the secondary structures got specific forms. Those structures are visible with bare eyes on mountains or Earth's crust. However, primary structures are observable only through microscopes.

Secondary structures of rocks

Geologists did observe that the rocks in some parts of the continent have two morphological structures: *folded* and *fractured*, as shown by the sketches in Figure 1.12.

As mentioned above, various stresses (i.e., force/area) do create the secondary structures in rocks. According to the law of physics, a force acts between two bodies in that one exerts it, and the other experiences it.

The falling off the apple, in the story of Isaac Newton, was led by the gravitational force. The force was exerted by the Earth and experienced by the apple.

Again, according to Newton's third law of motion, *every action has an equal and opposite reaction.*

So, when Earth pulled the apple downward, the apple also pulled the Earth upward. But, in reality, nobody did observe the Earth moved upward. Why? Because the upward acceleration of the Earth was negligible compared to the downward acceleration of the apple. According to the second law of motion,

Force = Mass × Acceleration.

So, the acceleration is inversely proportional to the mass of the body.

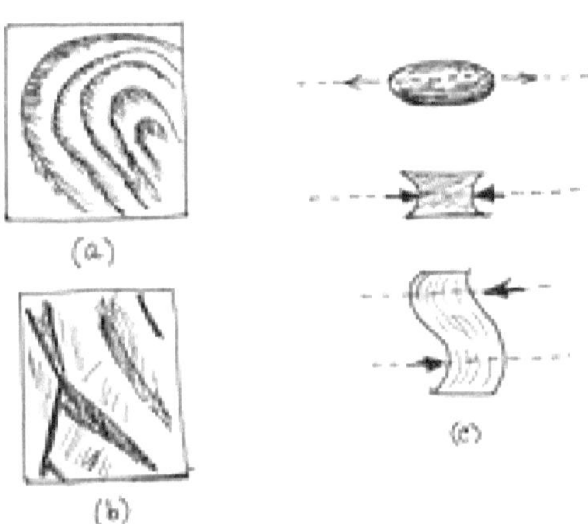

Figure 1.12. The secondary structures of rocks: (a) folded, (b) brittle, and (c) deformed. The deformed structures as occurred due to tensile stress, compressive stress, and shear stress, from top to bottom, respectively.

The secondary structures of rocks get formed due to external forces. The deformation of rocks occurs along the direction of the applied force, which causes either temporary or permanent structural modifications.

Imagine a rubber ball. If you press, it gets deformed but restores the shape due to the elasticity. However, a plastic ball may or may not regain its original form depending on the magnitude of the force applied to it. Rocks are like plastic bodies.

Question: *Are the deforming forces fundamental?*

Our knowledge: The fundamental forces need the exchange

of virtual particles to avoid collision and annihilate. It is a *non-classical* concept.

Answer: In the case of classical push-pull forces, the bodies apparently come in contact. Thus, those forces appear *non-fundamental*. However, at the microscopic level, push or pull gets generated by the electrostatic interaction between the electrons in the two bodies. And, at the level of 'electron-electron' interaction, there will be a minute gap between the two bodies of the classical domain as well while implementing action and reaction of the force. It implies that the deforming forces are also of fundamental type.

DEFORMING STRESSES: Stress is a force per unit area, which can be of three types: *tensile, compressive,* and *shear.* Tensile stress elongates, while compressive stress compresses a body such as rock due to coaxially pulling or pushing, respectively, on the opposite sides. However, shear stress causes relative displacements of different layers of a body due to the pulling or pushing along non-coaxial lines.

Figure 1.12 illustrates secondary structures of rocks: (a) folded, (b) brittle, and (c) deformed. The deformed structures as occurred due to *tensile stress, compressive stress,* and *shear stress,* shown from top to bottom, respectively.

When two tectonic plates collide, tremendous shocks, i.e., pressure waves, develop, which do not spread uniformly in all directions through the lithosphere. Consequently, it creates an immense deformation, mostly folding in the Earth's crust.

The folded structures in rocks get visualised due to the loss of

horizontality in the strata. The folding of the rock is an endogenic process. There are two types of folding: *anticlines* and *synclines*. Anticline forms ridge by folding upward, and syncline forms trough by folding downward. The folding of rocks usually occurs in a pair of anticline and syncline processes as we can see it in the mountains. When rocks deform plastically, folding appears. On the other hand, brittle bring about fractures, which occur in well-defined planes or zones, and sometimes it can form a deep fissure or crevice in the rock.

Primary structures of rocks

There are three different primary rocks: *igneous*, *sedimentary*, and *metamorphic*.

Molten substances underneath the Earth's crust remain under very high pressure and temperature. If there is an egress, the molten materials flow out of the Earth's surface, creating volcanic eruptions. On cooling and solidifying at a rapid rate, those ejected materials become extrusive (volcanic) igneous rocks. However, if the cooling and solidifying processes take a long time, intrusive (plutonic) igneous rocks do form, usually, with crystalline structures. The size of the crystalline rocks depends on how fast or slow was the cooling. In the case of fast cooling, small crystals grow, while large crystals grow in slow cooling.

The process of sedimentation of rocks is quite different from that of the igneous rocks. Igneous rocks are born hot, in contrast to sedimentary rocks that are inbred cold at the Earth's surface. Examples of sedimentary rocks are quartz and clays, which do form by the physical breakdown and chemical alteration of rocks in the flow of water or wind. Sand is composed of quartz, and mud is of clay minerals. Over time, when sand or clay gets packed underneath the Earth's crust due to high pressure and temperature (<100 °C), sandstones and shales do form, respectively.

The high pressure at any depth arises due to the weight of the upper portion of the Earth. It is an effect of the downward gravitational force towards the centre.

Shells and corals are also sedimentary rocks. The earlier ones get formed from the dissolved calcium carbonate, or silica, in seawater, and the later ones from the dead plants buried deep under the Earth's surface. There are many sedimentary rocks, which need not discuss here. Sedimentary rocks may undergo further modifications underneath the ground. For example, mineralisation of dolomite into limestone, and the formation of petroleum, coals, and ores.

The metamorphic rocks do form when existing rocks get exposed to high pressure and temperature, or hot mineral-rich fluids deep inside the Earth where tectonic plates meet. In the process of metamorphism, rocks do not melt but become denser. Examples of some common metamorphic rocks are schist, marble, and gneiss.

With the help of modern experimental techniques, scientists can analyse the structure, physical and chemical properties of solids, which have two distinct morphologies: *amorphous* and *crystalline.*

Structure of solids

There are so many varieties of solids around us. Let us consider two of them: glass and quartz. Both get moulded from silica (SiO_2), which is also the constituent of sand and is one of the most abundant materials on Earth. It does exist in several minerals and synthetic products. It is a chemical compound of silicon and oxygen atoms that are conjugated by the covalent bonds, as shown in Figure 1.13.

The covalent bond is an electron sharing interaction between two or more atoms. In SiO_2, there are two oxygen atoms (O_8^{16}), each has eight electrons: two in the innermost K orbit and six in the outermost L orbit. *According to the octave rule, the last shell of an atom can have up to eight electrons.* So, an oxygen atom can accept an additional two electrons in the L orbit. A silicon atom has fourteen electrons: two in K orbit, eight in L orbit, and the remaining four in the outermost M orbit. Therefore, it can share its four electrons with other atoms and form a bond as though a "give-and-take" relationship, and thus, forms a compound like SiO_2. Figure 1.13 (a) shows the formation of covalent bonds between Si and O atoms, and Figure 1.13 (b), the symbolic representation of this compound.

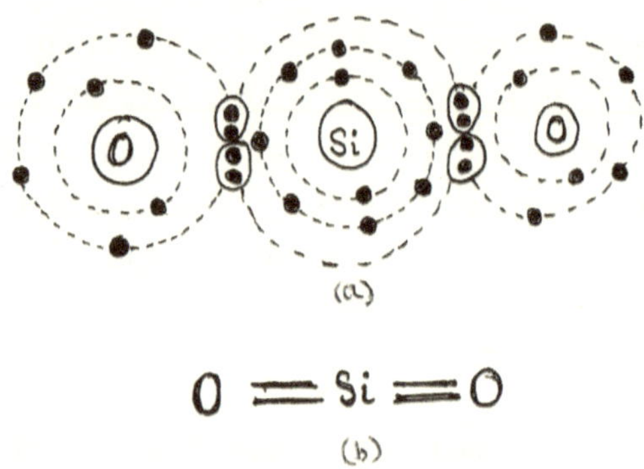

Figure 1.13. Silica compound

By sharing of electrons, both Si and O fulfill the octave rule in their respective last shells. Due to stronger electronegativity, oxygen pulls the electron pairs in the bond towards itself. Consequently, both Si and O become polarised with a small positive charge ($2\delta+$) at Si-side and small negatives charge ($\delta-$) at each O-side. It leads to an electrostatic interaction between the atoms in nearby molecules of SiO_2. In solids, it is the arrangement of constituent atoms or molecules through bonds, which give them specific structures.

What is the structural difference between glass and quartz? The earlier one is an amorphous solid, whereas the latter one is crystalline. The word *amorphous* has originated from two Greek words *a* (means, *without*) and *morphé* (means, *shape*, or *form*). Thus, an amorphous is solid with no long-range ordering of atoms or ions, in contrast a crystal.

In crystals, the arrangement of atoms or ions is such that the separation between them in the lattice is the same everywhere. It is called *periodicity*. In a periodic lattice[5], any atom or ion can get located using a well-defined translational vector from any other location of atom or ion in the lattice. The number of nearest-neighbours, called the *coordination number*, in a crystalline lattice remains the same at all lattice points. But it is different at different points in an amorphous solid. Figure 1.14 shows cartoons of the two-dimensional arrangements of atoms in (a) amorphous solid and (b) crystalline solid where '*a*' and '*b*' denote the periodicity in the two directions of the lattice.

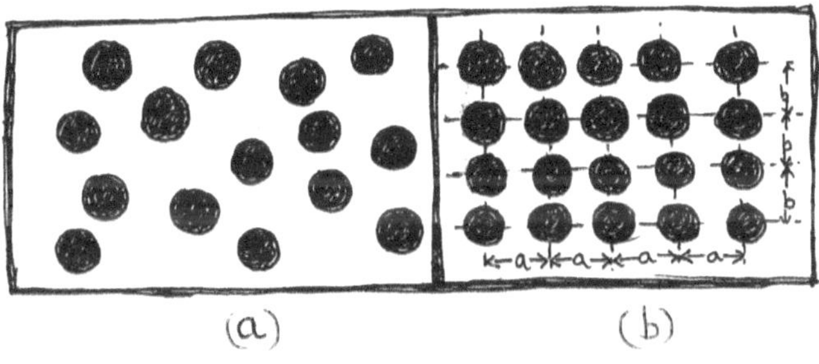

Figure 1.14. Atomic arrangements in (a) amorphous solid and (b) crystalline solid.

SOLIDIFICATION HISTORY MATTERS. A melt can solidify either by slow or fast cooling. The rapid cooling, such as from the melting temperature directly to a low temperature, is called the *quenching*. We know that anything happening at a rapid rate has less, or no, control on the system. The same does happen in the arrangement of atoms (or ions) while solidifying a material by quenching.

In the molten state, atoms or ions have a random distribution. In a quenching, the system does not get time to control the dynamics of the molten state. Thus, atomic or ionic arrangement gets locked at random positions producing an amorphous solid. When the melt is cooled slowly, atoms lose the thermal energy steadily and get arranged in a periodic distribution creating a crystalline solid of minimum verve. The same does happen while solidifying a melt of SiO_2 into a glass or quartz.

Which congealed is more stable, and why? With common sense, we can say it is the ordered solids, i.e., crystals, because of a minimum energy configuration. It can be compared analogically with the traffic condition on a busy road during peak hours. When vehicles run in order, no traffic jam occurs, and a stable situation remains. However, if those vehicles run randomly, a traffic jam develops that leads to an unstable road situation.

Interactions in solids

Why atoms or ions arrange in an ordered pattern in crystals? Because an interatomic force drives them in a specific motif to minimise the potential energy of the whole system. The nature of that interactive force depends on a few inherent physical properties such as size and charge of the building blocks, i.e., atoms or ions.

For a crystalline solid, the potential energy has a well-defined form, which is not so for amorphous solids. The interatomic vibrancy of a crystalline solid has both attractive and repulsive components. While the earlier one restrains atoms moving away

from each other, the other one acts only when two of them come in the closest proximity to prohibit a collision. The standard model of the interatomic potential energy in crystalline solids is as given by the *Lennard-Jones equation* (1924),

$$U_{12} = 4\varepsilon \left[\left(\frac{\sigma}{r}\right)^{12} - \left(\frac{\sigma}{r}\right)^{6} \right] \tag{1.6}$$

where ε is a constant and 4ε defines the least verve for a regular lattice, σ denotes an interatomic separation with which the potential energy of the lattice becomes zero (i.e., for $r = \sigma$, $U_{12} = 0$). r represents an arbitrary separation between the atoms; for a crystal, the interatomic separation (say r_0) becomes such that the lattice energy reaches a global minimum. That interatomic separation dictates the periodicity of the lattice. The shape of the Lennard-Jones potential energy has shown in Figure 1.15 (a). There are a few other energy models, which I have skipped.

The potential energy of a crystalline solid is known as its *cohesive energy*. Cohesion has several meanings, such as the action or fact of forming a united whole. In the following section, I shall discuss the driving forces between the atoms in solids that give rise to the cohesive energy. Amorphous solids, on the contrary, have no single minima, like crystalline solids, in the potential energy profile. Either it has several minima, as shown in Figure 1.15 (b), or no clear minima. Amorphous and crystalline solids can be differentiated as follows:

In crystalline solids atoms, ions or molecules arrange in an ordered manner but in amorphous solids, they arrange in a disordered manner; crystalline solids do have definite three dimensional (3D) shapes but amorphous solids do not have any well-defined 3D shape; crystalline solids do have sharp melting temperatures but amorphous solids do have broad melting temperatures; crystalline solids do show all properties of solids and they are called 'true solids' but amorphous solids do not show all properties of solids and they are called 'pseudo solids'; energy in crystalline solids is lower than that of non-crystalline solids; a crystal is a stable solid but an amorphous is a metastable solid—thus if an amorphous solid is kept at a temperature just below its melting point for a long time, its constituent atoms, ions, or molecules can gradually rearrange themselves into a more ordered fashion and the solid can transform into a crystalline solid.

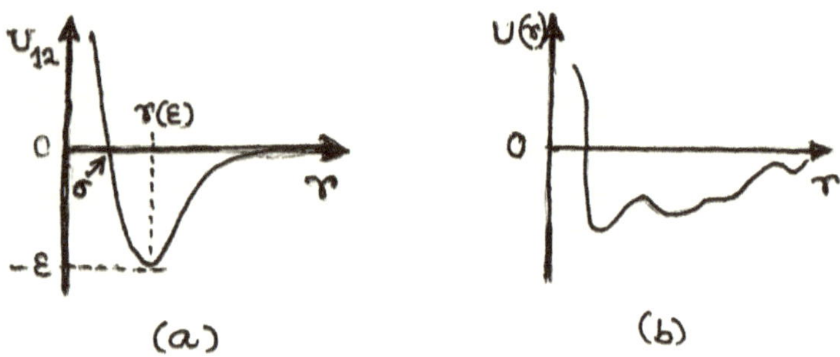

Figure 1.15. (a) Sketch of the Lennard-Jones potential; r(ε) represents the minimum stable distance between two atoms below which there is repulsion and above which there is attraction, and (b) sketch of the interatomic potential energy of a glassy system; it has several minima corresponding to several short-range ordering.

DIVERSITY AMONG CRYSTALS: There are four types of pair interactions in crystalline solids: *van der Waals, ionic, metallic,* and *covalent.* Thus, there are four types of crystals: *van der Waals crystals* (e.g., solid argon), *ionic crystals* (e.g., NaCl), *metallic crystals* (e.g., Na), and *covalent crystals* (e.g., diamond).

MORE DETAILED PICTURES: Van der Waals crystals form of neutral atoms, where there is no "give-and-take" bonding. Ideal gas atoms helium (He), neon (Ne), argon (Ar), krypton (Kr), xenon (Xe), and radon (Rn) are the examples. In such crystals, the attractive force arises due to the *induced dipole-dipole interaction* between the atoms.

A dipole gets induced to a lattice atom due to an anisotropic Coulombic compulsion from the nearest neighbour atoms. Such dipole gets configured with a negative pole of electrons' gathering at one side, and a positive pole of the nucleus at the other side, as shown by the cartoon in Figure 1.16. The dipole-dipole attraction becomes dominating when atoms move away from each other, while the repulsion rules at the nearest proximity between the constitutional units. The repulsive interaction follows the *Pauli Exclusion Principle.*

Pauli Exclusion Principle: *No two electrons in an atom or molecule can have the same four quantum numbers (n, l, m_l, and m_s). As an orbital can accommodate a maximum of two electrons, the two electrons must have opposing spins.*

Figure 1.16. Pictorial representation of an atomic dipole (ellipsoidal) with '+' pole representing its nucleus and '−' pole representing a cluster of electrons.

An ionic crystal contains ions, instead of neutral atoms, in the long-range motif (called *ionic lattice*) connected through the electrostatic interaction. Examples are the alkali halides, say sodium chloride (NaCl), as shown in Figure 1.17. In metallic crystals, such as iron (Fe), metal atoms do form the lattice through metallic bonds. The melting temperature of metals varies widely depending on the bond strength.

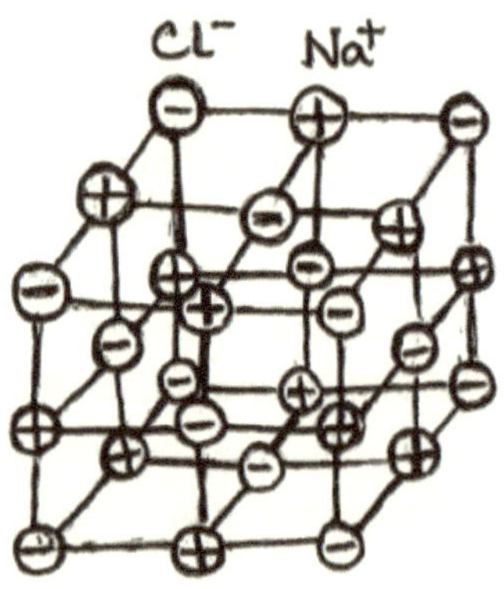

Figure 1.17. Pictorial representation of ionic crystal NaCl.

In metallic crystals, atoms are so close that the electrons in their outermost orbits experience equal pulls from their parent nuclei as well as neighbouring nuclei. Consequently, those electrons become delocalised or free to move randomly in the lattice. That makes each atom a positive ion with free-electrons around. Those free electrons are responsible for the electricity in our homes. Such a positive massive ion and free-electron combination, as illustrated in Figure 1.18, creates the metallic bond and an attractive electrostatic potential in the lattice.

In covalent crystals, the lattice of atoms forms due to the covalent bonding, as explained for SiO_2 (Figure 1.13). The covalent bond is the strongest among all four types that make covalent crystals non-malleable and hard. For example, diamond is so hard that it can cut glass. Graphite, though another covalent crystal of carbon, is soft. The difference arises due to their distribution of bonds in the two crystals, as shown in Figure 1.19. The diamond has a three-dimensional (3D) network of covalent bonds; in contrast, graphite has covalent bonds in the two-dimensional (2D) planes that fabricate a layered structure. Thus, graphite layers are as hard as the diamond, but the interlayer bonds are weak van der Waals type. Consequently, graphite layers slip relative to each other when an in-plane pressure has imparted, which makes it a soft and brittle material.

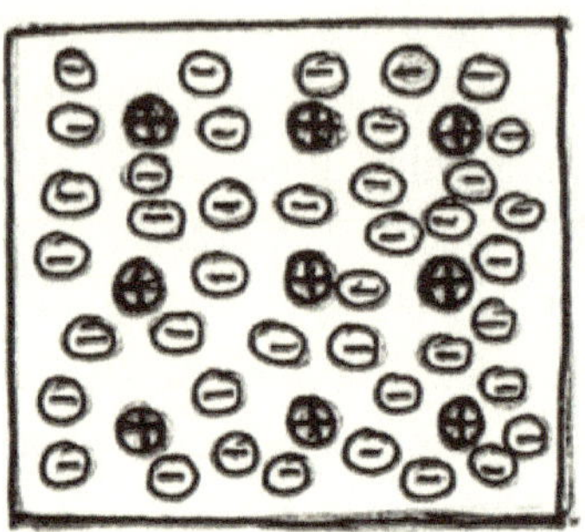

Figure 1.18. Sketch of metallic bond; (+) represents lattice ions and (-) represents 'free electron' forming cloud around ions mediating electrostatic interaction.

BONDS CONTOL PHYSICAL PROPERTIES OF CRYSTALS: Many physical properties of solids depend on their bond strength (see Table 1.1). Strength-wise bonds have the order as *covalent bond > ionic bond > metallic bond > van der Waals bond.*

As the metallic bond is weak, the metallic solids are malleable. On the other hand, the van der Waals bond is the weakest one, which is unable to form a solid at ambient temperature and pressure. Usually, van der Waals crystals form of molecules, and their typical melting temperature is less than or near 0 °C. Thus, many molecular crystals remain as liquid (e.g., water) or gaseous (e.g., oxygen) at room temperature.

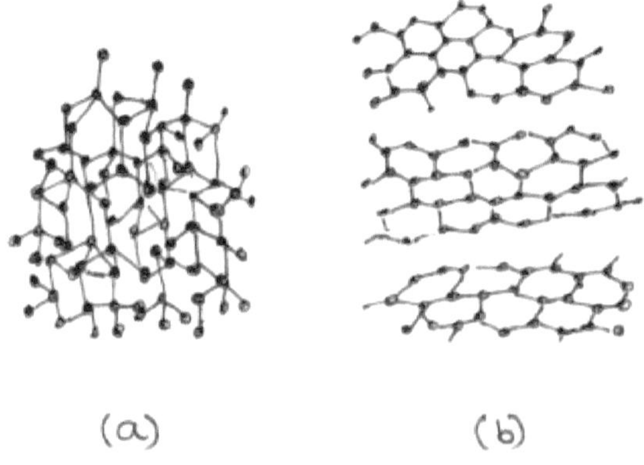

Figure 1.19. Covalently bonded carbon (C) atoms in (a) Diamond crystal, and (b) Graphite layered structure.

Physical properties of solids

Table 1.1. Few physical properties of solids based on their lattice bonds.

Solids	Physical properties
Covalent	As the bond is directional, physical properties can be highly directional or anisotropic, i.e., vary along with directions. Covalent solids have a high melting point and usually hard. Those solids are electrical insulators, except graphite, in which each carbon has one free electron which can conduct through interlayer spacing.
Ionic	This bond is non-directional; physical properties are also isotropic, i.e., the same in all directions. Ionic solids are brittle along the cleavage planes (e.g., planes containing ions). Those solids are poor electrical conductors, and their hardness depends on their ionic charges.
Metallic	Metallic bonds are highly symmetric; the physical properties of metals are always isotropic. Those solids are highly electrical and thermal conductors due to the availability of free electrons, usually soft and have low melting points, and malleable.
Van der Waals	It is the weakest bond and unstable, as the induced dipole-dipole bonds form and break spontaneously, which makes the crystal very fragile. Noble gas atoms form this kind of crystal structure, which can easily break into gas state. Melting and boiling temperatures of those solids are very low, usually less than 0 °C.

ELECTRICAL PROPERTIES: The interatomic bonds define the electrical properties of solids, as *conductors, semiconductors,* and *insulators.* Bohr's atomic model proves that the electronic orbital energy has an inverse proportional relation with their radii (r). It implies that the binding energy of electrons in the innermost orbit is the highest, while in the outermost orbit is the lowest. Thus, escaping from the atom is facile for the remotest electrons.

Figure 1.20 depicts a typical picture of the energy (in eV) states of different orbits (i.e., n = 1, 2, 3…) in an isolated atom; for the innermost one (K orbit, or K shell) n = 1, and the subsequent by n = 2, 3, 4, and so on. Those energy levels suggest that an equivalent or more amount of energy will be required to remove an electron from those respective orbits of a neutral atom. For example, hydrogen has only one electron in the K orbit; to remove that and ionise the atom, 13.6 eV energy will be required.

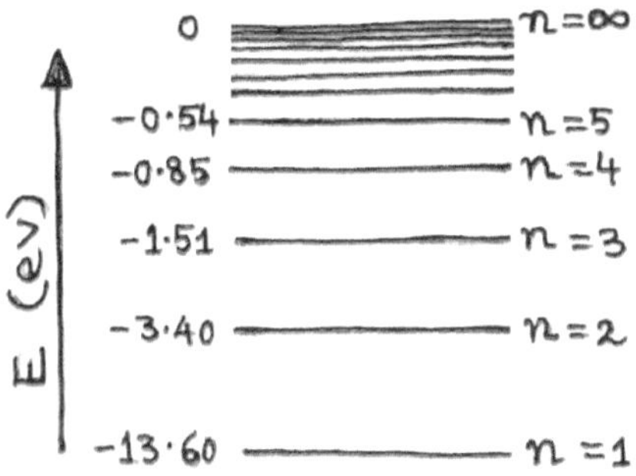

Figure 1.20. Energy levels (in eV) of different Bohr orbits (n = 1, 2, 3…) in an atom; n = 1 is the innermost orbit.

In crystals, atoms in the lattice produce a periodic electric field, which is known as the *crystal field*. This field interacts with the electronic states of atoms and modifies their energies in the lattice. Thus, each energy level spreads into a band in the crystal following the *Pauli Exclusion Principle*, as shown in Figure 1.21.

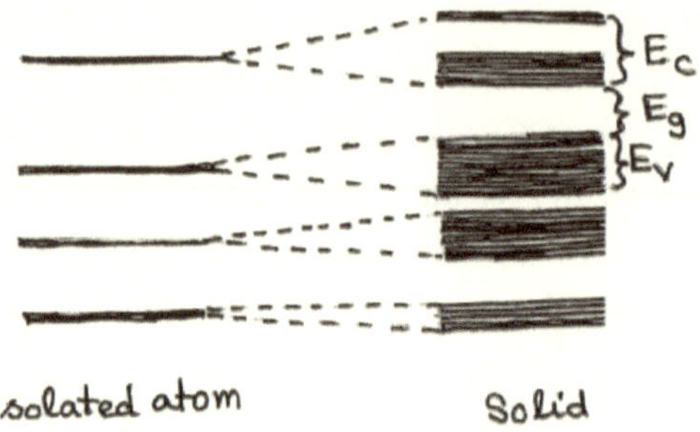

Figure 1.21. Electronic energy band formation in solid from isolated atoms.

The splitting of the electronic energy states in the presence of an electric field is known as the *Stark effect*. The name came after the German physicist Johannes Stark, who won the Nobel Prize in Physics in 1919 for this discovery.

The low or inner energy bands in a lattice remain fully occupied by the electrons. However, the highest energy band remains partially filled by the valence electrons; thus, known as the *valence band* (E_V). Next, there is a *conduction band* (E_C), which may have partially filled by the free electrons like in conductors. There is a gap between the highest energy state of the E_V band and the lowest energy state of the E_C band, which is called the *bandgap* (E_g), where no electron can exist. Thus, a supply of energy is required for electrons to jump from the E_V band to the E_C band. Depending on the width of E_g, solids are classified into three types, as described below.

For conductors (such as copper and aluminum), E_g is zero; i.e., the valence band partially overlaps with the conduction

band. Consequently, few electrons spontaneously move from the valence band to the conduction band. Upon application of an external voltage across a conductor, more electrons move to the conduction band, which produces a large electric current.

For semiconductors, $0 < E_g \leq 3$ eV. For example, silicon and germanium have the gaps 1.1 eV and 0.7 eV, respectively, a small voltage (but higher than that for the conductor) can generate a finite current.

Insulators (e.g., glass, plastic, and wood), in contrast, have a much wider bandgap, around 9 eV. Thus, produce no measurable current even with a very high voltage. Insulators are usually amorphous solids.

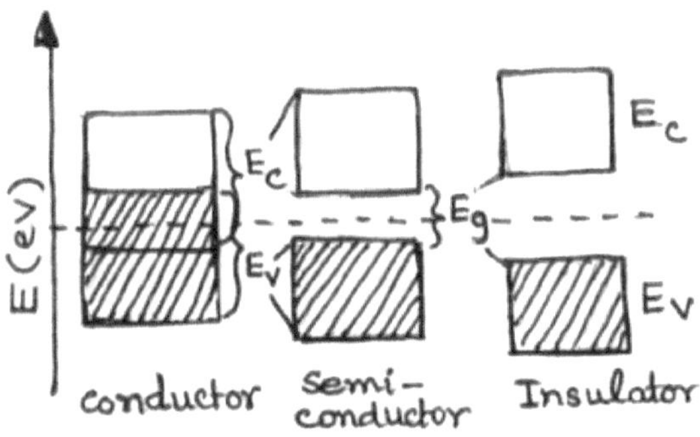

Figure 1.22. Schematic representations of energy band structures of conductor, semiconductor, and insulator.

Figure 1.22 shows the band diagrams for conductor, semiconductor, and insulator. The dashed line at the centre of the energy gap is called the *Fermi level*, and the energy of this level is called the *Fermi energy*, E_F. The Fermi level is the highest filled energy level of electrons in metals at absolute zero (0 K) temperature.

Metals proclaim high electrical conductivity, $\sigma \sim 10^3 - 10^8$ S.cm^{-1}, semiconductors low conductivity, $\sigma \sim 10^{-8} - 10^3$ S.cm^{-1}, and insulators almost no conductivity $\sigma \sim 10^{-18} - 10^{-8}$ S.cm^{-1}.

There are many applications of conductors, semiconductors, and insulators. Our ultramodern civilization is dependent on electricity. Without a conducting circuit, it cannot reach our home for running the air-conditioner, refrigerator, television, and many other electrical appliances. However, for safety, conducting wires are coated with insulating materials. Solar cell semiconductors also produce electricity from sunlight. Semiconductors are now becoming an alternative source of power.

So, we can realise how electromagnetic force instigates different physical properties in matter. Another example is the magnetic properties of solids. Magnetic materials also have many applications such as contrast agents in magnetic resonance imaging (MRI), in computer hard disk, in recording media, and magnetic levitation (MagLev).

When there is electric current through a conductor, it produces a magnetic field. As mentioned earlier, *a static charge generates an electric field, and a moving one a magnetic field.*

MAGNETISM ORIGINATES WITHIN THE ATOM: An electron, in an atomic orbit, executes an orbital motion around the nucleus and a spin motion around its body axis. Those

motions produce tiny current loops and, in turn, magnetic fields. Each electron in a Bohr orbit has, thus, two magnetic moments, i.e., *orbital* and *spin magnetic moments*. The spin magnetic moment is usually small and neglected. Classically, the orbital magnetic moment of electrons is $\mu=IA$ where I is the orbital current ($= e/T$, e = electronic charge and T = time-period of orbital rotation of electrons), and A ($= \pi r^2$) is the area subtended by the orbit. Thus, the orbital moment of the outermost electrons in an atom contributes maximum to the magnetism in a material. Or, the surface electrons of atoms define the magnetic property of a matter.

(a) (b)

Figure 1.23. Schematic representations of magnetic orientations of electrons in orbits with: (a) even number of electrons and (b) an odd number of electrons.

How does magnetism vary from one to the other material? Each electron behaves like a tiny magnet having north and south poles due to their spin. According to the Pauli Exclusion Principle, two electrons in the same energy state have opposite spins. Thus, if an atom has an even number of electrons in the outermost orbit, usually all get paired with up-down spins, as shown in Figure 1.23 (a), resulting in zero magnetic moments.

If an atom has an odd number of electrons in the outermost orbit, one (or more) electron(s) remain unpaired, as shown in Figure 1.23 (b), resulting in a net magnetic moment. As seen from Figure 1.3, hydrogen has only one electron in its K orbit, and thus, has a net magnetic moment whereas helium has a pair of electrons and no net magnetic moment. When an atom with a net magnetic moment (μ) gets exposed to an external magnetic field (B), its spin tries to align in a direction with the applied field, so that the energy of the system, $U(\theta) = -\mu B cos\theta$, where θ is the angle between $\vec{\mu}$, and $\vec{B}$, becomes minimum. The parallel alignment of spins with $\vec{B}$ makes θ equal to zero, and U (= $-\mu B$) a minimum. This property of the hydrogen atom is useful for magnetic resonance imaging of organisms in the human body (Gulkowski and Olchowik, 2006). Therefore, magnetism is an outcome of the spin interactions of the electrons in atoms, which is called *magnetic interaction*, and it is a component of the electromagnetic interaction. Depending on the magnetic interaction, magnetic materials can be ferromagnetic, paramagnetic, diamagnetic, or antiferromagnetic. These details are available in any graduate physics textbook.

OPTICAL PROPERTIES: Some materials show interesting properties upon interaction with light. In general, there are two types of materials, i.e., *transparent*, which allow light to transmit through, and *opaque*, which restrain it from passing within. Transparent materials such as glass and gems may appear either colourless or coloured. For example, emerald appears green, amethyst as violet, aquamarine as marine blue, and ruby as red. These colours appear when sunlight containing seven colours (VIBGYOR) passes through a gem, say ruby, its other hues get absorbed, but the red gets transmitted. When we see ruby due

to the transmitted light, it appears transparent red. Instead of sunlight, if a blue light incident on ruby, it would appear black because the cent percent blue light will get absorbed. An unstained glass looks colourless because the cent percent sunlight transmits through it. In contrast, an opaque material either fully absorbs or reflects, or both, the sunlight; consequently, it appears either black, colourless, or hazy coloured. For the same reason, flowers and leaves appear so colourful. Non-crystalline materials, except glass, are usually opaque.

The optical properties of matter depend on their interaction with light photons. The colour of light refers to its frequency. As mentioned above, sunlight has seven colous, i.e., rays of seven frequencies. If one of those frequencies matches the vibrational frequency of atoms in a material, it gets absorbed. The remaining frequencies get transmitted or reflected. Consequently, that material appears coloured. Usually, no material absorbs all colours of the sunlight except the *black body*, which is non-existent in reality.

The reflection of light could be either regular due to a perfectly flat surface like a mirror, or irregular due to a rough surface. A mirror is a perfect reflector, i.e., it reflects all colours of white light. Thus, we can see our image on the mirror the same (except the lateral inversion) as our original body. The irregular reflection is called *scattering,* and surface roughness of the order of a micron can cause it. If a body scatters all wavelengths of the sunlight, it would appear white. The sky does appear blue as that colour from sunlight gets scattered by the tiny dust particles (called *aerosols*) at very high altitudes in the atmosphere. During the rise or set of the sun, a part of the sky appears pink or orange. It is because of the scattering of both red and yellow colours of the sunlight. The vast ocean appears blue due to the reflection of blue light, coming from the sky, by the ocean surface. However, the foams in the wave appear white as the sunlight gets cent percent scattered by the vigorous wave surface.

In crystalline solids, the unabsorbed light transmits through the material. Because of the transmission, we can see through the spectacle. Colour filters in camera work in transmission mode. While propagating from one medium (say, air) into another (say, glass), the path of light deviates from the original. This phenomenon is called *refraction*, and the capacity of deviating light is the *refractive index* of a material. Thus, the angle of deviation is proportional to the refractive index of the material.

Light is an electromagnetic wave of a wide range of wavelengths. For example, the wavelength range of visible lights is 4000–7000 Å. The electric field of an electromagnetic wave interacts with the electrons in every matter resulting in their various optical properties as discussed above. Therefore, the light-matter interaction is electromagnetic, which has many applications. For instance, the splitting of sunlight using a prism, known as *dispersion*. The formation of a rainbow, or a holographic image, is an example of the disbandment of light. Magnification of an object using a lens, determination of chemical compositions of material through absorption, or transmission of light using a spectroscopic technique are the examples of light-matter interaction.

Other electromagnetic waves such as x-rays, gamma, or

ultraviolet rays (in the high-frequency range), and infrared rays (in the low-frequency range), also participate in electromagnetic interactions with matter, which help in various applications for the welfare of human being. For example, x-rays-matter interaction helps in the determination of crystal structures using the x-ray diffraction (XRD) technique. In crystallographic applications, the wavelength of the x-rays (λ) varies in the range of 1–2 Å, which are called *hard* (high energy) x-rays. For medical diagnoses such as imaging of bone fracture, the x-rays used have wavelengths in the range of 10–100 Å, which are called *soft* (low energy) x-rays. Our tissues are transparent to x-rays, but bones are opaque that makes it possible to get the image of a bone fracture within our body.

There are many examples of light-matter interactions in nature. Those interactions are beneficial for the lives on Earth. When sunlight falls on the green leaves of a plant, it does photosynthesis to produce starch through the electromagnetic interaction between sunlight and a pigment called *chlorophyll*. Sunlight contains high-frequency ultraviolet (UV) rays, which can damage our skin by breaking the bonds in skin proteins. So, to protect the skin, sunscreen lotion is used that reflects the UV light. An ozone layer in the Earth's atmosphere also does protect us from UV light. When the sunlight passes through the ozone layer, ozone molecules (O_3) absorb its UV frequency and break down as oxygen molecules (O_2): $2O_3 \xrightarrow{UV} 3O_2$ through the electromagnetic interaction.

There are numerous varieties of nonliving materials on Earth. All of them undergo various interactions that are driven by the four fundamental forces. In the following sections, I shall

discuss the living organisms. But living organisms can't form without a favourable environment, which includes water.

Water on Earth

How did Earth accumulate so much water? There are many hypothetical answers to this question. The *dehydration melting hypothesis* says that a large quantity of water came from various internal materials of Earth, e.g., leakage of water stored in hydrated minerals of rocks (Schmandt et al., 2014). The *extraplanetary sources hypothesis* says, the collision of comets, trans-Neptunian objects, or water-rich meteoroids (from asteroid belts) may have brought water on Earth (Daly et al., 2018). Let us avoid those details and discuss the structure and properties of water.

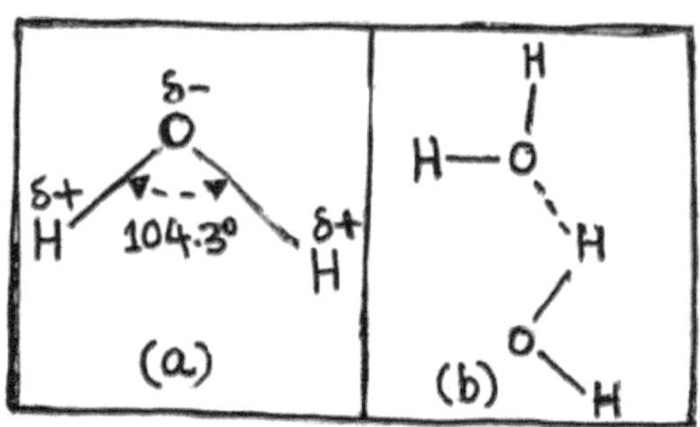

Figure 1.24. Schematic representations of (a) water (H_2O) molecule, and (b) association of water molecules through H-bonding

STRUCTURE OF WATER: A water molecule (H_2O) is composed of two hydrogens (H) atoms and one oxygen (O) atom through covalent bonding, as shown in Figure 1.24 (a). The angle between the two covalent bonds is 104.3°, i.e., the water molecule has a bent structure, which makes it polar. The oxygen atom in H_2O is electronegative, which tugs the electron-pairs in covalent bonds towards itself. Consequently, the O-end becomes a negative pole ($2\delta-$), and the two H-ends become positive poles ($\delta+$). But this does not make the H_2O molecule bend. There is another force behind this. Remember that an O-atom has six electrons in its outermost L shell, two get paired with each other, and two make the covalent bonds with two H atoms. The remaining two unpaired electrons in the L orbit repel each other, forcing the two covalent bonds to contract their angle from 180° to 104.3°.

PROPERTIES OF WATER: Due to its polar character, water possesses several unusual properties, such as (i) it forms weak electrostatic *hydrogen bond* (*H-bond*s) between nearby H_2O molecules, dashed line in Figure 1.24 (b), causing high surface tension in water, (ii) it is a super-solvent; it carries ions and molecules in our body for metabolism, (iii) it has higher boiling temperatures compared to other liquids, (iv) it has a high dielectric constant, which gives it a very high hydrating capacity, and (v) it has high heat capacity, which helps it retaining heat for a longer time compared to other liquids. H-bonds form due to electrostatic attraction between the H-atom of one H_2O molecule and the O-atom of a nearby one. It is a very flexible bond and easily breaks in the liquid state of water, but somewhat stable and stretched in the ice state. Thus, the density of ice is less than that of water causing ice to float on water. Water is an essential component of lives on Earth. About seventy percent of Earth's surface has filled with water; a human body also has almost an equal percentage of water.

Atmosphere around Earth

WE ARE IN A LUCKY ZONE! So far, there is no evidence of life in other planets in our solar system. Why? One reason is temperature. The Earth got a place at the *Goldilocks zone* or *habitable zone* in our solar system, where the thermal condition is just fitting for maintaining enough water in the liquid form. The average temperature of Earth is about 16 °C, which is apt for the survival of life as well. Mercury and Venus are much closer to the sun; therefore, too hot to hold water in liquid form. Mars, Jupiter, Saturn, Uranus, and Neptune are, on the other hand, away from that lucky zone; therefore, too cold to melt ice (if available) into water. Besides, to hold life, there must be hard surfaces like rocks, but Jupiter, Saturn, and Uranus are gaseous planets. The atmosphere also must contain necessary gases such as carbon dioxide (CO_2) and ozone (O_3) to keep the terrene warm and protected from harmful radiation. CO_2 also helps in photosynthesis by the plants. Both animals and plants require oxygen (O_2) for respiration. In our solar system, Earth is the only planet where all those necessary ingredients for life are present. Also, the geomagnetic field protects lives from the cosmic background radiation.

Today our atmosphere contains about seventy-eight percent of nitrogen, which is one of the fundamental components of proteins, and proteins are the building blocks of living cells. Besides, several other gases such as oxygen (28%), argon, carbon dioxide (CO_2), helium, neon, krypton, hydrogen, and ozone, and many other components enrich our atmosphere. Earth has a vast ocean that suffices many needs of life. None of those existed in the

beginning but acquired through various evolution of Earth.

As mentioned earlier, hydrogen and helium were the main ingredients of Earth's environment in the beginning, along with dust, which could not fulfill the needs of life. With time those gases also disappeared. As the Earth cooled, the gaseous atmosphere started developing again and retained until today due to strong gravitational attraction. Several reports indicated that the life-supporting atmosphere and oceans got formed on Earth about 4.5 billion years ago. The reorientations and collisions between the plate tectonics had caused several earthquakes that brought out volcanic eruptions of melted metals, gases, and water vapour. Those erupted materials helped to establish the necessary atmosphere around Earth and the formation of crust on it. There is another hypothesis that says when the Earth's crust had cooled down sufficiently, a large amount of the vapor content in the atmosphere condensed as water. Thus, ponds, lakes, and oceans formed on Earth. The presence of water made the Earth further apt for life. However, for a long time, the atmosphere was devoid of oxygen. How did oxygen appear on Earth? We shall know it in the next chapter.

In the beginning, life grew only under the ocean, and those were only single-cell organisms. Slowly, more complex multi-cellular organisms evolved through a long journey, known as *the evolution of life*. Why did it take so long time? It is because, as Charles Darwin (Darwin, 1859) claimed, evolutionary transformation occurs through natural selection. How does natural selection occur? It happens through interaction with the surroundings, including the environment, nonliving materials, and other living organisms. Life has evolved to higher-order animals and to the human being, called *Homo sapiens*. Evolution flows only in one direction; the American theoretical physicists, Sean Michael Carrol said,

"I'm trying to understand cosmology, why the Big Bang had the properties it did. And it's interesting to think that connects directly to our kitchens and how we can make eggs, how we can remember one direction of time, why causes precede effects, why we are born young and grow older. It's all because of entropy increasing."

At the end of this chapter, let us get back the three questions asked in the The Beginning and assess how many we can answer now.

1. *What was the attractive force in the pre-universe particle?*

Ans. We have realised that the attractive force in the pre-universe particle was a unified form of all fundamental forces. So, this question needs no further query.

2. *Does that force exist now?*

Ans. Yes, that unified force exists now at every nook and corner of this universe, but in four different forms. Those are the forces, which caused the evolutions of all animate and inanimate objects. This question also has got resolved.

3. *What is the identity of God?*

Ans. Still not known. We shall look for this answer at the end of our journey in the Conclusion.

CHAPTER TWO

EVOLUTION OF LIFE

PURPOSE OF THIS CHAPTER: To know how do fundamental forces participate in creating life on Earth.

The age of life on Earth is estimated based on the dating of rocks that contain fossils, or by looking at the 'molecular clock' (i.e., mutation rate) of DNA in living organisms. The modern genetic analysis also provides molecular-level information on the process and the time of the evolution of different species from a single lineage. This chapter has been prepared based on the following questions:

a) *How did atoms form the building block of living organisms?*

b) *What were the first living organisms on Earth?*

c) *How did organisms evolve from simple to complex structures?*

d) *What is the 'tree of life'?*

The building block of living organisms

ORIGIN OF LIFE ON EARTH: There are two hypotheses on the inception of life on Earth:

(i)   The *biogenesis hypothesis* claims that life has originated from the existing-life

(ii) The *abiogenesis hypothesis* predicts that it was from nonliving matter.

THEORY SUPPORTED ABIOGENESIS: In the 1930s, Soviet biochemist Alexander Oparin and British-Indian scientist J.B.S. Haldane independently published their research in support of the abiogenesis hypothesis using a *primordial soup theory*.

PRIMORDIAL SOUP THEORY: This theory says, in the presence of energy, e.g., ultraviolet radiation from the sun, or lightning discharges in clouds, or heat released in volcanic eruptions, molecules such as CO_2, N_2, H_2, H_2S, and CO (except O_2), in the primordial atmosphere of Earth, reacted together and formed some elementary organic compounds.

EXPERIMENT SUPPORTED ABIOGENESIS: To verify the primordial soup theory, Stanley Miller and Harold Urey of the University of Chicago, USA, experimented in 1952, known as the *Miller-Urey experiment*. They applied an electric discharge to a gaseous mixture of vapor, CH_4, NH_3, and H_2 in a sealed vial for a week. The reaction had produced several organic molecules in the solution, including amino acids, approving the primordial soup theory. Subsequently, the abiogenesis hypothesis has been

accepted as the origin of life on Earth.

AMAZING CARBON: Carbon is a crucial element for living organisms; the reason is the following. A carbon atom has four electrons in the outermost L shell making its valency four. Thus, it can form covalent bonds with four other elements like hydrogen (H), oxygen (O), nitrogen (N), sulfur (S), phosphorus (P), as well as with itself. It can form single, double, or triple covalent bonds, chains, branched, and ring molecules. This capacity makes it the most versatile and unique element comparing other elements on Earth. Thus, an abundance of carbon helped to form organic molecules and initiate life on Earth.

MOLECULES OF LIFE: The compositions and functions of four major organic molecules, i.e., *carbohydrates*, *lipids*, *proteins*, and *nucleic acids*, are stated in Table 2.1. Those organic molecules form through covalent bonds due to electrostatic interaction and are highly stable.

In the beginning, the prime source of carbon in the Earth's atmosphere was carbon dioxide (CO_2), which is a poison for life. Carbon gets continuously recycled between the atmosphere, plants, and animals, known as the *carbon cycle* of life, through several processes like photosynthesis, respiration, combustion, and decomposition.

What is life? Life is a differentiating factor between the nonliving and the living objects. The living body grows up through the cell division. Bending of plants towards the sunlight, or fear of danger in animals gets stimulated by some physiological chemicals such as hormones, which are absent in nonliving bodies.

Table 2.1. Compositions and bio-functionalities of four major organic compounds.

Organic compounds	Examples	Compositions	Functions
Carbohydrates	sugars, starches	carbon, hydrogen, oxygen	Provides energy to cells, stores energy, forms body structures.
Lipids	fats, oils	carbon, hydrogen, oxygen	Stores energy, forms cell membranes, carries messages.
Proteins	enzymes, antibodies	carbon, hydrogen, oxygen, nitrogen, sulfur	Helps cells keep their shape, makes up muscles, speeds up chemical reactions, and carries messages and materials.
Nucleic Acids	DNA, RNA	carbon, hydrogen, oxygen, nitrogen, phosphorus	Contains instructions for proteins, passes instructions from parents to offspring, helps make proteins.

Animals do physical as well as mental activities. Human beings can think, imagine, and do many other activities that are related to the mind. Life has a soul, energy! Nonliving matter also have cohesive energy or chemical energy, but not a soul. The soul disappears as soon as a human being dies, but the chemical energy remains in the body. So, what is the soul? According to Encyclopaedia Britannica, the soul is the incorporeal essence of a living being. In Sanskrit, it is called *atman*, which defines the identity of an individual esse. Thus, the soul is the identity of a person, which disappears from the body after death. According to the German biophysical chemist, Manfred Eigen (1996),

"Life is not an inherent property of matter. Life is indeed associated with matter, but it appears only under very specific conditions and, when it does, it expresses itself in very diverse and individual ways."

First organisms on Earth: prokaryotes

Shreds of evidence from fossil indicate that the first appearance of life on Earth was about 3.8 billion years ago; it was under the sea in the forms of *bacteria* and *archaea*. At that time, the Earth's atmosphere was devoid of oxygen. Thus, life grew up with anaerobic processes, and those organisms were named *prokaryotes*.

How did the prokaryotes form? As mentioned above, first, a group of organic molecules got formed under the sea through the abiogenesis process. Some of those were capable of self-replicating. Around 1960, scientists discovered that the self-replicating modicum was a *ribonucleic acid* (RNA) molecule.

An RNA is the composition of a sugar molecule, a base of nitrogen, and a chain of phosphate (PO_4^{3-}) groups. All are attached by the covalent bonds, as shown in Figure 2.1 (a). There are four types of nitrogen bases: adenine (A), uracil (U), guanine (G), and cytosine (C). Those are the covalently bonded rings of carbon, hydrogen, nitrogen, and oxygen atoms. An RNA molecule has a spiral chain structure, as shown in Figure 2.1 (b), with a backbone of phosphate chain. Sugar molecules and bases get projected from the phosphate chain. In the prebiotic stage of Earth, RNA did play the dual role, i.e., as the carrier of genetic information and as the foldable molecule. The RNA-protein or biomolecule interaction is a very significant physiological activity, which depends on RNA folding.

Figure 2.1. Schematic representations of RNA as (a) molecule, (b) single strand.

How does an RNA fold? There are several *in vivo* factors as it happens now within a living body. However, those factors were absent in the primordial Earth. So, we can only guess an answer as follows. In an RNA molecule, oxygen, nitrogen, and hydrogen atoms make covalent bonds. However, in the covalent bonds, the electronegative oxygen and nitrogen atoms pull the electron-pairs near to them, creating local dipoles in the molecule. Consequently, different segments of the RNA molecule could fold due to electrostatic attractions.

Through electrostatic interactions, RNA-protein complexes had formed first while blooming of life in the aqueous environment. RNA, being hydrophobic (water-hating), was encapsulated by the proteins to avoid the aquatic medium. Bacteria, archaea, and other prokaryotes were those RNA-protein complexes in the Primordial Earth. Thus, according to the Harvard chemist Walter Gilbert

(1986), it was *the RNA world*. But according to Linus Pauling,

"Life is a relationship among molecules and not the property of one molecule"

The abiogenesis hypothesis aims to investigate *how chemical reactions gave rise to life in the matter* (Voet and Voet, 2004), as life gets realised in specific chemistry of carbon and water. It depends mainly upon four types of organic molecules (Ward and Kirschvink, 2015), listed in Table 2.1.

In higher organisms, DNA, and not RNA, carries genetic information. DNA is *deoxyribonucleic acid*. It is another nucleotide molecule like RNA, but more complex and heftier. Scientists do believe that unlike RNA, DNA might not form spontaneously in Earth's prebiotic environment. Probably, the RNA-based life had switched to DNA-based life, as it works better at storing information. But according to Christopher Switzer of the University of California,

"Organisms could have used naturally occurring DNA, then developed the tools to make their own"

Matthew Powner of the University College of London suggested,

"Life may have begun not as the RNA world, but as RNA and DNA world in which both were intermingled."

A DNA molecule has one phosphate group, one sugar group, and four nitrogen bases: adenine (A), thymine (T), guanine (G), and cytosine (C). The fundamental dissimilarities between RNA and DNA include:

- RNA has a ribose sugar group, whereas DNA has a deoxyribose sugar group.

- RNA is a single-stranded (ss) molecule, whereas DNA is double-stranded (ds) molecule.

- RNA remains in the nucleus, cytoplasm, and ribosome, whereas DNA the nucleus and mitochondria in a body cell.

- RNA gets synthesised by DNA in the body cell when needed, but DNA is self-replicating.

However, in primordial Earth, RNA was self-replicating.

Figure 2.2 shows the molecular skeleton (a) and the folded structure of ds-DNA (b). Each sugar ring makes a covalent bond with a nucleotide base. A and G bases are known as *purine*. Each of them contains a hexagonal ring and a pentagonal ring, as shown in Figure 2.3. C and T bases are known as *pyrimidine*. Each pyrimidine gas only a hexagonal ring. A and T bind through two H-bonds, while C and G bind through three H-bonds.

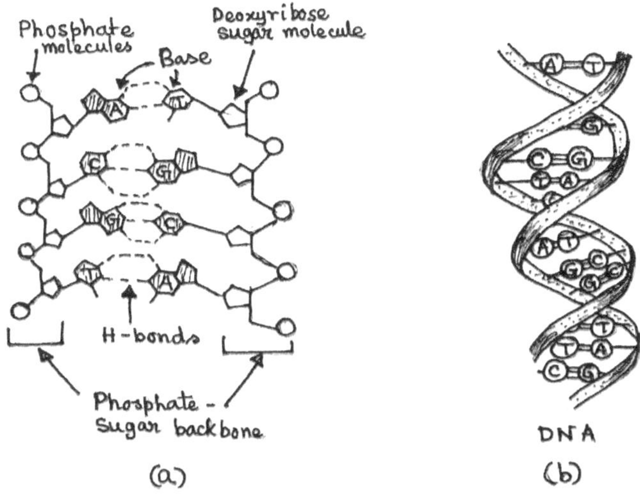

Figure 2.2. Pictorial representations of DNA: (a) molecular constitution, (b) double-strand folded structure.

In DNA base pairing, nitrogen (N) and oxygen (O) atoms do form weak H-bonds with the hydrogen atoms that hold the two strands together. However, H-bonds break easily, and DNA strands open up. That helps in the transcription of DNA (copying of DNA to RNA) and replication (copying DNA to DNA); the two essential processes for duplicating cells in cell division. Both RNA and DNA consist of hydroxyl groups (−OH), but an RNA has four, whereas DNA has three such groups. That makes DNA *deoxy*. Since RNA gets copied from a portion of DNA, it has a smaller length.

Figure 2.3. Pictorial representation of DNA base pairing through hydrogen bonds.

The structure of a double-helix DNA was discovered by James Watson, Francis Crick and Maurice Wilkins in 1953, and they were awarded Nobel Prize in Physiology in 1963. Rosalind Franklin was also an active contributor in that discovery. She died of breast cancer in 1958 at an age of 37 years.

INTRA-DNA INTERACTIONS: DNA needs a specific folding to perform specific physiological functions. Such folding is a consequence of different interactions such as pairing and stacking of bases. The base pairing occurs as described earlier. The stacking involves *hydrophobic* and *electrostatic interactions* depending on the aromaticity[6] and dipole moments of the base rings. The strength of interactions increases with the concentration of salt (i.e., ions) around bases. The presence of ions masks the repulsion between the negatively charged phosphodiester backbones. The third interaction that helps base-stacking within the same strand is van der Waals type, which arises due to the formation of temporary local dipoles.

How does a hydrophobic interaction develop? Non-polar molecules such as fatty acid (i.e., oil) have a phobia to polar solvents like water and avoid interacting with them. It is called the *hydrophobicity*. In the presence of a hydrophobe, the H-bonds between water molecules break to make space for them without any one-to-one interaction. It is an endothermic process. The distorted water molecules then make new H-bonds to form an ice-like structure, called a *clathrate cage*, around the hydrophobe. That forces hydrophobe to compact in a reduced volume. Thus, a drop of oil in water takes a spherical shape of the smallest surface area.

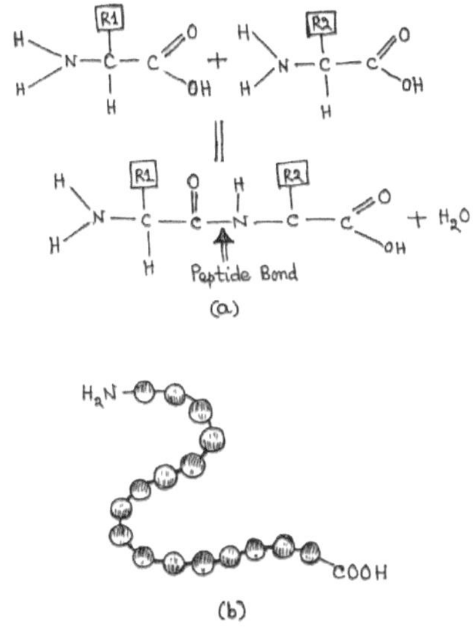

Figure 2.4. (a) Peptide bond formation between two amino acids and release of one water molecule, R1, and R2 side chains specific to the amino acid, (b) protein's primary structure (polypeptide chain) with one end containing an amino group and the other end a carboxyl group; circles represent different amino acid residues.

Proteins are polypeptide chains of amino acid residues with one side containing an amino group ($-NH_2$) and the other side carboxyl group ($-COOH$). Amino acids do arrange in specific sequences through peptide bonds, which gives the *primary structure* to a protein. A peptide bond is a covalent bond between the $-OH$ group of one amino acid and the $-NH_2$ group of the next one. A water molecule (H_2O) gets released during the formation of a peptide bond, as illustrated in Figure 2.4 (a). Thus, it is known as the *dehydration* or *condensation* process.

The amino acid sequence in a protein, as shown in Figure 2.4 (b), stores specific information, which has described by the *theory of Conservation of Information*. The size and molecular weight of a protein are determined based on the number of residues in its primary structure. It also gives the structure-property relationship of proteins (Hatton and Warr, 2015). Proteins have three types of folded structures, called *secondary*, *tertiary*, and *quaternary structures*, which do form upon several interactions, similar to those described above.

Oxygen in Earth's atmosphere

ANAEROBIC ORGANISMS: Due to the lack of oxygen in the primordial atmosphere on Earth, the metabolic process in unicellular prokaryotic organisms, such as archaea and bacteria, was anaerobic. They used hydrogen, sulfur, and other chemicals to harvest energy through chemical reactions that generated carbon, sulfur, and nitrogen cycles on Earth. About 3.2 billion

years ago, those prokaryotes evolved with a green pigment called *chlorophyll*. They are known as *cyanobacteria*, who used CO_2 and water to produce food (i.e., glucose) in the presence of sunlight, and release oxygen (O_2). This process, as we know now, was *photosynthesis*. Chlorophyll helps to trap sunlight to ignite the photosynthesis reaction as follows:

$$6CO_2 + 12H_2O \xrightarrow{sunlight} C_6H_{12}O_6(glucose) + 6O_2 + 6H_2O$$

Sunlight breaks the covalent bonds of CO_2 and H_2O to cause a chemical reaction with the help of chlorophyll to form glucose, oxygen, and water. Glucose is a big molecule with more number of covalent bonds compared to the reactant molecules that make it a *storeroom of chemical energy*. Thus, the photosynthesis process transforms the light (or heat) energy into chemical energy.

Thus, the atmosphere was gaining O_2 that caused the evolution of oxygen-based life on Earth. As the atmosphere and ocean became increasingly oxygenated, organisms that used oxygen spread and eventually came to dominate the biosphere. At present, the Earth's atmosphere contains around twenty-one percent oxygen.

Advanced organisms: eukaryotes

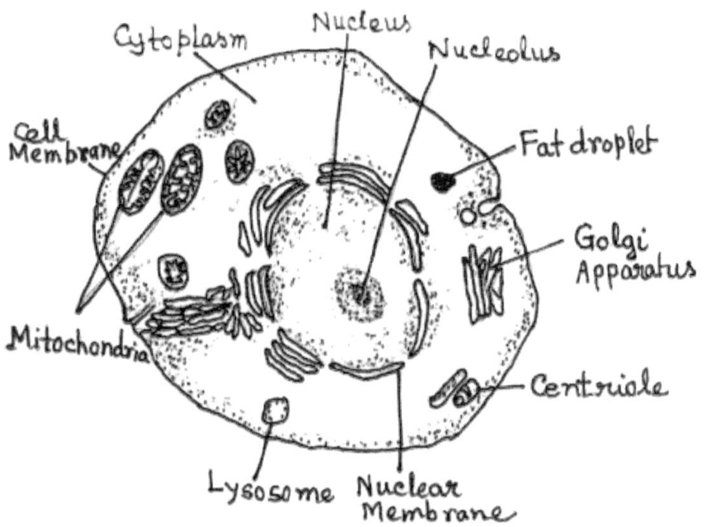

Figure 2.5. Sketch of a eukaryotic cell.

EVOLUTION TO AEROBIC ORGANISMS: Accumulation of oxygen made the atmosphere poisonous to anaerobic bacteria. For survival, they had to alter their metabolic processes that led to the appearance of aerobic organisms about 2.5 billion years ago. That evolution brought many changes in the structures and functions of many biomolecules. For example, RNA changed from the circularly folded self-replicating form to the linearly folded non-replicating molecule.

The first eukaryotic cells evolved about 1.5–2 billion years ago. Such an advanced clique contains one nucleus and several organelles embedded within a jelly-like medium,

called *cytoplasm* that contains water, salts, and proteins. Thus, it gives the cell shape and protection to its nucleus and organelles. The entire unit has enveloped with the protein-lipid double-layer membrane. There is a cytoskeleton, i.e., a network of proteins, spread over the cytoplasm to provide rigidity to the cell structure. Figure 2.5 shows the sketch of a eukaryotic cell. Table 2.2 has a list of few organelles and their respective functions.

The nucleus is also filled with a jelly-like medium, called *nucleoplasm*, which protects its contents such as nucleolus and chromatins of DNA, RNA, and proteins that carry genetic information from parents. There is a porous membrane covering the entire unit.

Table 2.2. List of few organelles and their functions in a eukaryotic cell.

Organelle	Function
Nucleus	DNA storage
Mitochondria	Energy production
Smooth Endoplasmic Reticulum (SER)	Lipid production, detoxification
Rough Endoplasmic Reticulum (RER)	Protein production (for export out of the cell)
Golgi apparatus	Protein modification and export
Peroxisome	Lipid destruction (contains oxidative enzymes)
Lysosome	Protein destruction

How did eukaryotic cells evolve from prokaryotic cells? The *Endosymbiosis theory*, proposed by the Boston University biologist Lynn Margulis in the 1960s (Martin*et al.*, 2015), has the answer. It explains that small prokaryotes got entrapped inside large organisms and created symbiotic associations. Those associations evolved as eukaryotic cells.

What is endosymbiosis? 'Symbiosis' means an interactive association of two or more different organisms that give them mutual benefits, and 'endo' means within. Thus, endosymbiosis is an intracellular 'give-and-take' relationship.

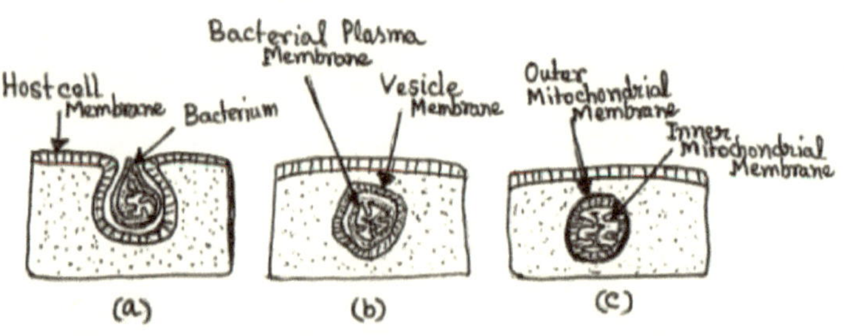

Figure 2.6. Sketches show the endocytosis process in 'symbiotic partnership'.

It was difficult for the anaerobic bacteria to sustain in the oxygen-rich environment. Therefore, they ingested aerobic bacteria so that their parasites could absorb the environmental oxygen for their metabolism and save hosts from fatal toxicity. In return, the aerobic symbiote could earn food that was ingested by the host bacteria. In that way, they maintained a symbiotic relationship. Figure 2.6 shows the endocytosis (i.e., engulfing) process of developing a symbiotic partnership.

Lynn Margulis showed that the incorporation of one bacterium into the other did produce many cell organelles. Mitochondria and chloroplast are two such organelles. Their circular DNA molecules resemble those in prokaryotic cells. Moreover, their size, membrane structure, and reproduction process (i.e., binary fission) are like those of bacteria. Both mitochondria and chloroplast could inhale oxygen from the environment for metabolic activity, i.e., to produce energy for life through the following reaction (known as respiration):

$$C_6H_{12}O_6(glucose) + 6O_2 \rightarrow 6CO_2 + 6H_2O + energy$$

The other organelles that had evolved with different special functions lost their genes that were taken up by the evolving nucleus in the same organism. The sequences of those genes were like the nuclear DNA of the extant eukaryotes from yeast. In an article entitled, *Where the nucleus comes from*, Tom Misteli (2001) of NIH wrote:

"The analysis showed that nucleus-related genes such as the ones involved in transcription, replication, cell cycle, nuclear architecture or ribosome biogenesis shared homology with archaeal genes rather than with bacterial genes. On the other hand, yeast genes related to cytoplasmic functions, such as metabolism, stress response, detoxification or protein and ion transport, were more closely related to bacterial genes than to archaeal genes. This clear separation allows the conclusion that, in a eukaryotic cell, the nucleus is of archaeal origin, but the cytoplasm is of bacterial origin."

The endocytosis process brings polar objects such as food into the cell, as illustrated in Figure 2.6 (a). On sensing an element at the outer charged layer, the cell membrane forms an inward loop and engulfs the food forming a vesicle. Then the vesicle tears out of the membrane and goes into the cytoplasm for further action. The entire process is possible due to a unique membrane protein called *clathrin*. As eukaryotic cells were underdeveloped initially carried out endocytosis crudely. That caused them to undergo further evolutions.

Many eukaryotes continued secondary, or tertiary endocytosis processes by engulfing another eukaryote that already had undergone a primary or a secondary endosymbiosis, respectively, which had brought on Earth a considerable genetic diversity.

What was the first eukaryotic organism? In 1987, Cavalier-Smith claimed that the unicellular and anaerobic *Giardia lamblia* might be the first eukaryote and the missing link between prokaryotes and eukaryotes. Cavalier-Smith also claimed that Giardia had converted to the first aerobic eukaryote by ingesting anaerobic bacterium. The next generation of eukaryotes might have descended from that. Interestingly, Giardia lamblia had two identical nuclei, which is unusual.

Multicellular organisms

For the first three billion years, organisms were only unicellular. Those were bacteria (prokaryotes), algae, and protozoans (eukaryotes). "Unicellularity is clearly successful…,"

said Eric Libby, a theoretical biologist at the Santa Fe Institute in New Mexico. Yet, unicellular organisms evolved into multicellular form. It was probably more beneficial to *togetherness*. How did organisms realize it? Quoting from the book, entitled, *The Biology of Belief: Unleashing the Power of Consciousness, Matter, and Miracles*, by Bruce H. Lipton,

"Signal molecules used by individual cells to regulate their own physiologic functions, when released into the environment, also influence the behaviour of other organisms. Signal molecules released into the environment allow for a coordination of behaviour among a dispersed population of unicellular organisms. Secreting signal molecules into the environment enhanced the survival of single cells by providing them with the opportunity to live as a primitive community".

MAKE AND BREAK: Figure 2.7 illustrates the formation of a multicellular organism. Many organisms, afterward, gave up such a multicellular relationship and reversed back to unicellularity. For example, the bacterium Pseudomonas fluorescence rapidly evolved into a multicellular community to gain better access to oxygen but destroyed the coterie afterward. However, some organisms remained in the multicellular form, as described below.

Figure 2.7. The cartoon shows the organisation of unicellular organisms into a 'community'.

How did some organisms remain in multicellular forms? It was due to the *ratchet mechanism* that opposed the reversion to unicellularity (Libby *at al.*, 2016) from multicellularity. *A ratchet is a mechanical device that allows non-stop unidirectional movement.*

How did the multicellularism help? It helped to keep the integrity among the cells by interacting via cell membrane and extracellular matrix (ECM) and defend the microenvironment changes. The ECM is a composition of water, proteins, and polysaccharides (Frantz, 2010), which provides physical scaffolding for the cellular constituents, and initiates crucial biochemical and biomechanical cues for tissue morphogenesis, differentiation, and homeostasis. Each tissue has ECM with unique composition and topology.

MULTICELL TO TISSUE: Multicells did form either in an aggregative or in a clonal pathway (Grosberg and Strathmann, 2007). In the aggregative alley, cells need not have genetic equality, and some of them may cheat in collective activities that restrict evolution toward complex tissues and organs (Bonner, 2000). However, in the clonal process, coherent and conjugative interactions exist among the cells. That helped to generate a sense of togetherness and promote the formation of tissues.

Plants and animals have achieved multicellularity through the clonal pathway (King, 2004; Rokas, 2008). Single-cell (spore or zygote) divides into multiple cells, known as *cell division*, with identical genetic characters and cell-to-cell communication.

ELECTROSTATIC INTERACTION IN CELL DIVISION: In mammalians, there are two types of cell divisions, namely, *mitosis* and *meiosis*. In mitosis, a parent cell divides into two identical daughter cells to support the growth of organisms and limbs. Meiosis creates sex cells (eggs and sperms) that form the embryo of a new life and allows a genetic variation. The electrostatic interaction does regulate the process at different stages, such as condensation of chromosomes during the cell divisions (Maeshima et al., 2018).

CELL ADHESION MOLECULES: Cells need communication with each other to control the shape and functionality of tissues. The cell-to-cell adhesion is a selective process that is mediated by some glycoproteins, called *cell adhesion molecules* (CAMs). The interaction between the CAMs is either homophilic or heterophilic. Protein-protein and protein-carbohydrate interactions are examples. In the homophilic interactions, CAMs on one cell does interact with identical molecules on the other. Cadherin proteins work as CAMs in the presence of calcium ions (Ca^{2+}). The Ca^{2+} ions convert flexible cadherin molecules into stiff hooks, which link cells together. In the heterophilic interactions, CAMs in one cell-membrane behave as receptors that bind to specific molecules (known as *ligands*) in the other cell-membrane. For example, selectin proteins, in a cell membrane, have carbohydrate recognition domains that form covalent bonds with specific glycans (polysaccharide chains) in the other cell membranes. Integrins and immunoglobulin superfamily (IgSF) are also CAMs. Integrins are transmembrane glycoproteins. They bind the cytoskeleton of one cell to the ECMs of the other and monitor the adhesion. IgSF has the most diverse group of known receptors that are available in different species starting from insects to man. They have a common structural feature, the immunoglobulin homology domain. The above processes brought advanced life on Earth. Canadian author and science blog-owner, Jessa Gamble said:

"Life evolved under conditions of light and darkness, light and then darkness. And so plants and animals developed their own internal clocks so that they would be ready for those changes in light. Those are chemical clocks, and they're found in every known being that has two or more cells and in some that only have one cell."

Tree of life

The environmental progression triggered the unicellular to multicellular evolution. Similarly, the phenotype evolution, i.e., the change in individual genetic character, was also taken place due to the environmental change. Thus, the environment was a crucial factor for creating several species of plants and animals in the *tree of life*. English naturalist, geologist, and biologist, Charles Darwin developed it in 1859, based on the phylogenetic relationships among the creatures.

The tree, as shown in Figure 2.8, has three primary branches: *bacteria, archaea, and eukaryotes*. Each of these branches has several divisions, and further divisions, and so on. Subdivisions are called subgroups. For example, eukaryotes have several subgroups such as plants, animals, and fungi. Animals have subgroups such as sponges, cnidarians, and Bilateria (animals with bilateral symmetry); Bilateria has subgroups like arthropods,

molluscs, and nematodes, and it continues. Eukaryotes have more subgroups than archaea and bacteria. Each group or subgroup shares a common ancestry, called *Last Universal Common Ancestor* (LUCA).

Darwin had predicted that the species had evolved into several subgroups through the natural selection, i.e., through their continuous struggle (i.e., interactions) against all environmental adversities. He predicted it after several observations made on the finches, birds in the family of Fringillidae, in the Galapagos Islands (in the Republic of Ecuador), and many other species across the world. Quoting him:

"Owing to this struggle for life, any variation, however slight and from whatever cause proceeding, if it be in any degree profitable to an individual of any species, in its infinitely complex relations to other organic beings and to external nature, will tend to the preservation of that individual, and will generally be inherited by its offspring. The offspring, also, will thus have a better chance of surviving, for, of the many individuals of any species which are periodically born, but a small number can survive. I have called this principle, by which each slight variation, if useful, is preserved, by the term of Natural Selection, in order to mark its relation to man's power of selection."

Figure 2.8. The tree of life: main groups are bacteria, archaea and eukaryotes; each of which contain several subgroups.

ENVIRONMENTAL EFFECT ON EVOLUTION: Matter get exposed to the environment. So, there is always some interactions between the matter and the environment. Living organisms are no different. Our sense organs make a connection between our body with the surrounding. Therefore, the environmental changes directly affect our organ systems, cause *mutations* in the genes (the hereditary part of DNA). As genes are the carriers of a cell character, a transmutation brings changes in the cell structure and behaviour, which affect the tissue and the entire body. Those changes are sometimes useful for empowering the metabolic systems. Thus, there was a steady evolution of life on Earth.

The salient characteristics of DNA are to replicate itself in the newly born cells. In the cell, it is *transcribed* and *translated* before synthesising new proteins. Thus, a mutation in a gene due to the environmental factors creates an alteration in the protein sequencing, which may or may not be beneficial. Such molecular-level changes created macroevolutions of species and a great diversity of life on Earth.

DNA MUTATIONS: Mutation either alters or destroys a DNA base-pair sequence. There are two types of mutations: *point mutations* and *chromosomal aberrations*. In point mutation, more often, there will be a *point substitution*. Sometimes there could be a failure in copying one of the bases, known as *deletion*. There could be duplication or insertion of a different base, as illustrated in Figure 2.9. Cells can repair DNA. However, an imperfect repair leads to a mutation. Temporary mutations (*somatic mutations*) usually disappear, but permanent mutations (*germline mutations*) often bring steady changes over the coming generations. If there is a deleterious effect on the phenotype characteristics of an offspring, it brings a genetic disorder. A phenotype characteristic is an observable characteristic of an individual. However, a genotype is a set of genes that carries a particular trait.

Figure 2.9. Genetic mutations (genotype): substitution, insertion, and deletion.

If the mutation fits well in offspring, it is *an adaptation*. Such alteration can be an agent of evolution such as aquatic filamentous green algae to land plants, about 410 million years ago. Genomic mutation helps in the long-term survival of species, as it happens through natural selection and adaptation. A genome is a complete set of genes of a cell or an organism. In this regard, we should know the difference between *Darwinism* and *Lamarckism*. Lamarckism says the adaptation occurs through spontaneous interaction of organisms with the ever-changing environment; however, Darwin's adaptation means a sudden genetic alteration in an offspring that fits with the environment. Lamarckian adaptation leads to a slow continuous evolution based on the natural need of organisms; in contrast, Darwinian adaptation means a sudden evolution through mutation of DNA by chance. Surely, we know

that the evolution process is NOT SUDDEN, but CONTINUOUS INDEED! Lamarckian evolution is appearing to be scientific after the recent development of *epigenesis*, rather than Darwinian evolution.

TYPES OF EVOLUTION: There are three types of evolutions: *divergent, convergent,* and *parallel.* On a large scale, the tree of life represents the divergent evolution as closely related species had diversified into new habitats. In a small set, it means the evolution of humans and apes from a common ancestor. Divergent evolution has been affirmed by looking at the similarities in DNA among the species. Convergent evolution, on the other hand, takes place when species of different ancestry share similar traits because of much of a muchness environment. For example, whales and fishes have some common characteristics as both have evolved to swim through the water medium. Parallel evolution occurs when several species adapt independently, maintaining their same level of similarity or difference. It happens between unrelated species that do not have the same or similar niches in each habitat.

Many evolutionary biologists believe that the crossbreeding between species is a more sensible concept than the discrete evolutionary branches of the tree of life. The genetic examinations of bacteria, plants, and animals have proved such crossbreeding. So, it is not only that genes directly passing down individual branches, but there are transfers of genes between the species in different forks as well. That has resulted in a messier and tangled web of life. The crossbreed evolutionary concept puts a challenge to Darwin's theory of natural selection. Citing an article by Joel L. Carlin (2011):

"Most mutations occur at single points in a gene, changing perhaps a single protein, and thus could appear unimportant. For instance, genes control the structure and effectiveness of digestive enzymes in your (and all other vertebrate) salivary glands. At first glance, mutations to salivary enzymes might appear to have little potential for impacting survival. Yet it is precisely the accumulation of slight mutations to saliva that is responsible for snake venom and therefore much of snake evolution. Natural selection in some ancestral snakes has favored enzymes with increasingly more aggressive properties, but the mutations themselves have been random, creating different venoms in different groups of snakes. Snake venoms are actually a cocktail of different proteins with different effects, so genetically related species have a different mixture from other venomous snake families. The ancestors of sea snakes, coral snakes, and cobras (family Elapidae) evolved venom that attacks the nervous system while the venom of vipers (family Viperidae; including rattlesnakes and the bushmaster) acts upon the cardiovascular system. Both families have many different species that inherited a slight advantage in venom power from their ancestors, and as mutations accumulate the diversity of venoms and diversity of species increased over time."

The genomic mutation may also cause permanent extinction of species. A slow climatic change usually brings adaptation, while a drastic change may force extermination. When the Earth's atmosphere was abruptly changing with adversities, many species, such as dinosaur, could not adapt that changes and extinct. Examples of some declared extinct animal species are the Black Rhinoceros (West African), Chinese River Dolphin (China), Passenger Pigeon (North America), Tasmanian tiger (Australia), Dodo (Mauritius),

and many other species. Of late, global pollution and warming are of great concern, as those factors have adverse impacts on the ecosystem and may cause the extinction of a few animal species. Such global changes may even bring physiological and mental changes in human beings in the next generations.

Inbreeding depression, loss of genetic diversity, and mutation accumulation have hypothesised to increase disappearance risk (Frankham, 2005). According to Darwin's theory of *survival of the fittest*, only suited candidates are capable of modifying their metabolic system according to the climatic conditions and survive. Frankel (1970) was the first to propose that the loss of genetic diversity elevated the extinction risk, especially by compromising the evolutionary response to the environmental changes. Environmental factors such as temperature, light, and chemicals can cause *epigenetic* changes by altering the way molecules bind to DNA or by changing the structure of proteins that DNA wraps around. Consequently, an entire set of organisms in a body becomes unfit for fighting against environmental forces.

We are blessed to have extraordinary communicating skills with the environment and survived so long. The evolutionary history of human beings is very long, which can't get covered here. Briefly, the first human beings, known as *Homo erectus*, were living in Africa about 1–2 million years ago. Modern human beings, the *Homo sapiens*, arrived on Earth about 30,000 years ago. We are the subspecies of Homo sapiens, called *Homo sapiens sapiens*. Cartoons in Figure 2.10 show a famous picture of human evolution. Indeed, there were many changes in physical and brain structures. We expect to have more intelligent human beings in the future if there is adaptive evolution.

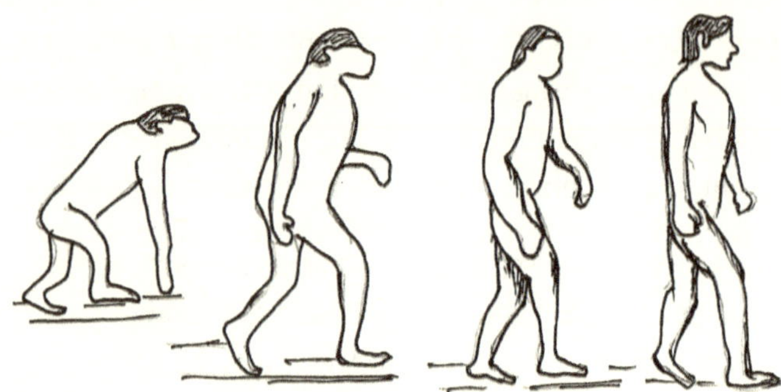

Figure 2.10. Cartoons representing the human evolution.

CHAPTER THREE

HUMAN PHYSIOLOGY AND MEDICINE

PURPOSE OF THIS CHAPTER: To know how do the fundamental forces assist in functioning of the organisms in a human body.

Science of the human body has two aspects, *anatomy* and *physiology*. Anatomy deals with the structure of organs, and physiology deals with their functions and inter-connections. The term *physiology* was introduced in the 1610s by the French physician, Jean Ferne. Probably, grabbed from the French word *Physiologie*, originated either from the Greek word *Phusiologia* or from the Latin word *Physiologia*. The meaning of Physiologia is the 'study of nature' as *physiologia = physio* (nature) + *logia* (learning). So, *Physiology is the learning of the natural functions of different organisms in a living body at its state of homeostasis.* Homeostasis means steady internal conditions. Physiology also concerns about medicines.

ATOM TO HUMAN BODY: Starting from the tiniest particles, a human body forms through the following bottom-up steps: *1) atoms (e.g., C, H, O, N) constritute biomolecules like proteins, RNA, and DNA through chemical bonding → 2) biomolecules combine to form fluid, organelles, and eventually cells → 3) similar cells interact together to form tissues (i.e., multicellularism) → 4) a set of those tissues combine into an organ → 5) two or more organs compose an organ system (e.g., digestive system) to work together → 6) many organ systems collectively form an entire human body.*

So, the above steps are the same as that of the synthesis of nonliving matter. However, a human body synthesis is an *in vivo* physicochemical process, whereas that of a nonliving matter is an *in vitro* chemical process.

Proteins are the building blocks of living organisms; synthesising proteins in cells is a sophisticated process. DNA (gene) contains the "blueprint" of making a specific protein in its nucleotide base sequences like ATCG. That specific sequence is translated into a protein sequence in two steps: *transcription* and *translation*; together, it is called *gene expression*, i.e., expressing the coded information in the gene of DNA into the new proteins.

CENTRAL DOGMA OF LIFE: Francis Crick, one of the discoverers of the ds-DNA structure, proposed that *the specific base sequences of DNA is transcribed into RNA and then the same sequences are translated into the new protein molecule.* During transcription, codes (of the specificity of a cell) get transferred from DNA to a type of RNA in the nucleus, called *messenger RNA* or *mRNA*. mRNA carries those codes from the core into the cytoplasm. During translation, mRNA messages those codes to the ribosome in the cytoplasm. Ribosome decodes those by reading the

base sequence of mRNA and synthesises amino acids accordingly. Each amino acid is composed of three bases. Each base is called a *codon*. There is another type of RNA, called *transfer RNA* or *tRNA*, which does the sequential assembling of amino acids, like a necklace, matching with the base sequence of mRNA. The sequencing process continues until the ribosome reads a 'stop' codon (i.e., a null code) in mRNA. So, DNA gets expressed like a carbon copy of a document. The mastermind behind the entire process is the brain. Such a flow of information from DNA to a protein is so significant that it is called the *central dogma of life*. "Dogma" means a belief without a scientific reasoning.

Thus, a cell forms protein and *vice-versa; which one comes first?* It is like the *hen and egg puzzle.* In primordial Earth, protein, RNA, DNA, or cells got formed through *in vitro* chemical reactions in nature. In the human body, a *zygote* forms within the mother's womb through breeding between a sperm cell (from the father) and an egg cell (from the mother), both made of specific proteins. Subsequently, the zygote divides (through mitosis) in two daughter cells, each having the same number and properties of chromosomes as the parent nucleus. Those two cells further divide, and it continues, which gives the birth of the *embryo* of a baby.

Each cell has four functions, namely, *cell division, cell growth, signaling,* and *metabolism.* Through the mitosis, a cell multiplies and forms an organism. Thus, cells are the building blocks of a living organism. The previous chapter has a brief narration of the cell division and cell growth. This chapter illustrates the signaling and metabolism processes in the human body.

Figure 3.1 shows the prime organs in the human body. An organ is a part of an organ system that can have functions integral to other organ systems as well. That way, organ systems form a network to perform the integrated metabolism in a multiplex body. Thus, the human body is a *holistic* system as described by Bruce H. Lipton in his book, *The Biology of Belief: Unleashing the Power of Consciousness, Matter, and Miracles.*

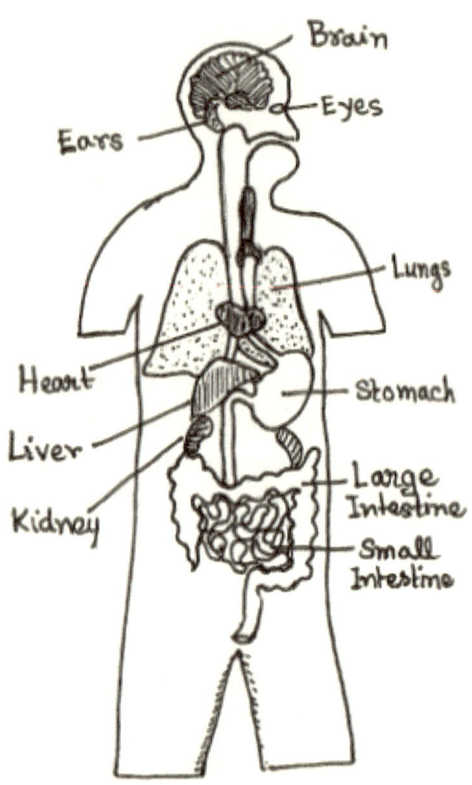

Figure 3.1. Schematic diagram of some major organs in the human body.

Nervous system: brain

The brain is the most complex organ in its structure and function. It has millions of nerve cells (called *neurons*), fibrous materials, and blood vessels. It has three vital parts: *brainstem* (lower portion), *limbic system and cerebellum* (middle portion), and *cerebral cortex* (upper portion), as shown in Figure 3.2. Environmental inputs and gene expression are the two factors that influence brain development. Though ninety percent of the brain develops within the first five years, the remaining takes many years.

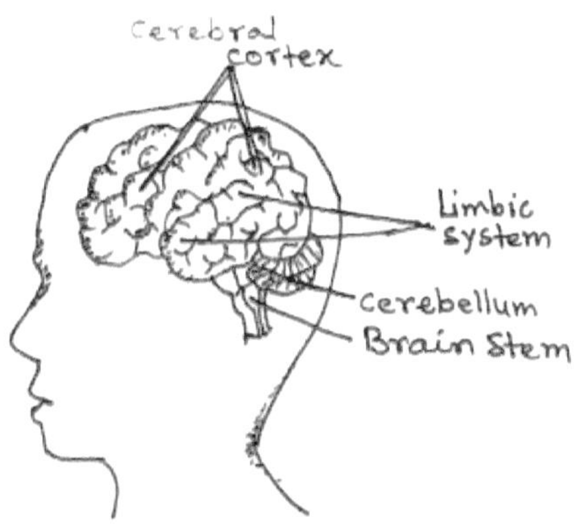

Figure 3.2. Schematic diagram of the human brain.`

The brainstem controls primary physiological activities such as respiration, heartbeat, blood pressure, and consciousness; the limbic system controls the expression of mental emotion and activities like eating and sleeping. The cerebral cortex is a significant segment of the brain, which is also known as the *cerebrum*. It is divided into two equal hemispheres, left and right. These hemispheres are joined together by a bundle of fibres, called *corpus callosum* that transmits messages from one side to the other. The left hemisphere, which controls the right side of the body, takes care of speech, comprehension, arithmetic, and writing activities. The right one, which governs the left side of the body, takes care of the creativity, spatial ability, artistic, and musical skills of a person. Each hemisphere has three lobes: *frontal, parietal* at the middle, and *occipital* at the back.

NEURON: A neuron has three components: *soma* (cell), *dendrites*, and an *axon*, as shown in Figure 3.3. The axon ends at capillary shaped *axon terminals*. Neurons connect each other via the dendrite terminals (for input) and the axon terminals (for output), and there is a small gap in between, called the *synapse*. There are four types of nerve cells: sensory neurons (receiving external inputs), interneurons (located in between other neurons), motoneurons (innervating muscle cells), and secretary neurons (triggering secretions). A nerve cell contains an electrolytic fluid, covered with a semi-permeable membrane. There is extracellular fluid. It has organelles like nucleus and mitochondria, as in body cells. The cell membrane is a double layer of lipid molecules, as described earlier for a eukaryotic cell.

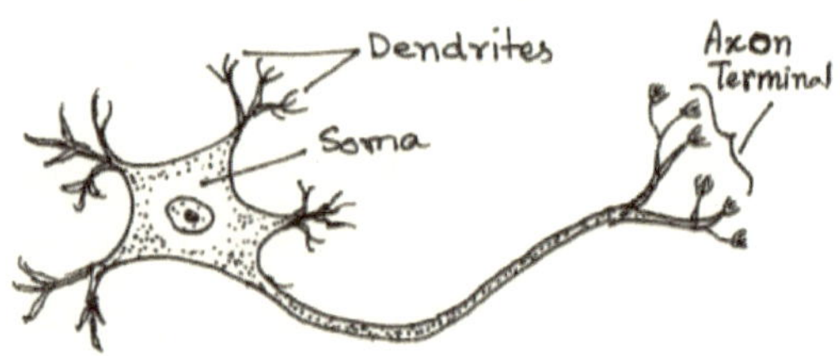

Figure 3.3. A simple representation of a neuron (nerve cell) structure.

Neuron membrane has ion channels through which small ions (e.g., Na^+, K^+, Ca^{2+}, Mg^{2+}, Cl^-) can pass through with different permeativities, but ions (e.g., organic) of bigger size are impermeant. Specific ions migrate through specific ion channels (e.g., Na^+/K^+ channel). These ions maintain a steady potential difference across the cell membrane, which is known as the *membrane potential*. The membrane potential may change due to changes in the permeativities of specific ions. In normal conditions, K^+ ions are more inside, and Na^+ ions are more outside. Both sides of the membrane have negative potentials, but the inner side has more than the outer side. If Na^+ ions enter the intracellular fluid, the charge on the inner membrane reduces. Consequently, the membrane potential also decreases. It is called *membrane depolarisation*.

FLOW OF INFORMATION: The depolarisation of the cell membrane brings the nerve cells into action. Changes in membrane permeativity can take place due to the interaction of a specific substance with membrane molecules. For example, in the retina cells (in eyes), the product of a photochemical reaction induces a change in ion permeativity and the membrane potential. The transient changes along the length from the soma to the axon terminals, with fixed duration and amplitude, are called *action potentials* or *nerve impulses*. Those are responsible for the transmission of information from one neuron to another (Oosting, 1979).

The spinal cord contains threadlike nerves. The brain-spinal cord combination is called the *central nervous system* (CNS). The nervous network that connects the CNS with the rest of the body is called the *peripheral nervous system* (PNS). The structural support, protection, and nourishment of nerve cells in the brain are governed by a set of cells called *ganglia* or *glia*. Glia is the plural form of glial; in Greek, it means glue. Those cells regulate the blood-brain barrier (BBB), allow nutrients and molecules to interact with the neurons, and insulate axons for an electrical signal to transmit faster.

Information passes from one part to the other in the brain through neurons via those synapses. When a piece of information reaches the terminals of a nerve cell, it stimulates pockets of chemicals, called *neurotransmitters*, and releases them into the synapses. Thus, neurotransmitters act as the bridges between the neurons. The activity of a neurotransmitter becomes very strong when we talk, think, enjoy, even when asleep. The brain functions depend on those neurotransmitter signal processing through neural networks.

BRAIN TROUBLE: If the neurotransmitter level falls, signal transmission disturbed, and that may cause mental disorders. Depression occurs if the corresponding neurotransmitter, i.e., *Dopamine* level falls (Chandrashekar 1987). The electrochemical signals in the brain are called *brain waves*. Brain waves can be mapped on a paper to know the mental condition of a patient by placing electrodes of a specially designed instrument on his scalp. This mapping of brain waves is known as an *electroencephalogram* (EEG). The antidepressant drugs such as Tricyclics, Monamine Oxidase inhibitors, or selective serotonin reuptake inhibitors (SSRIs) elevate the neurotransmitter level in the brain by preventing them from breaking down and getting reabsorbed into the neurons. It helps to relieve symptoms and

prevent the recurrence of the depression in a patient. Antidepressant drugs can improve mood, sleep, appetite, and mental concentration. Besides, those drugs improve *neuroplasticity*, i.e., the ability to form new connections between neurons in the brain.

BRAIN WORKS AS CPU: The brain communicates with the rest of the body through the CNS, and the twelve pairs of cranial nerves (CNs). Ten of that, coming out of the brainstem, control hearing, eye movement, facial sensations, taste, swallowing, and the face, neck, shoulder, and tongue movements. The remaining two pairs that are coming out of the cerebrum control smell and vision. When a stimulation arrives from the environment to a sensory organ (eye, ear, nose, tongue, or skin), it sends that to CNS via afferent (sensory) neurons of PNS, and then to the brain. The response from the brain and CNS reaches different parts of the body via efferent (motor) neurons of PNS. This way, the brain controls functions through the entire body, like the central processing unit (CPU) in a computer.

CPU is a microprocessor that receives input electrical signals, processes it based on the algorithms fed into it, and finally generates an output signal. It has a component called the *arithmetic logic unit* (ALU), which does logical computations similar to that of decision making in the brain. The human brain is many times complex than the CPU. Its decision-making process is not yet known.

Does the brain have ALU? In 2015, a study by a group from Okinawa Institute of Science and Technology (OIST) Graduate University in Japan discovered that there is a part in the prefrontal cortex, called the *striatum*, which makes the decision.

Striatum seems to work hierarchically with its three different sub-regions: ventral, dorsomedial, and dorsolateral, responsible for motivation, decision, and routine tasks, respectively (Bergland, 2015). Allying between the ventromedial and the dorsolateral decides from options. *The process involves many brain areas and interactivity among many neurons* (Montague, 2008). Scientists are trying to know the sinuous path from decision to action (Siegel et al., 2011).

Does the brain make all decisions? Not always. For example, if you touch a lukewarm hot-plate, the sensory nerves in that area of the skin instantly send a signal to the brain. After deciding the course of action, the brain returns the instruction to the muscles to pull the hand away. The entire cycle may take a few seconds. But, if the plate was too hot, action must be taken in a fraction of a second and can't wait for the brain's instruction. It requires a quick step, called the *reflex-action*. The neural pathway of the reflex-action is called the *reflex arc*, shown in Figure 3.4.

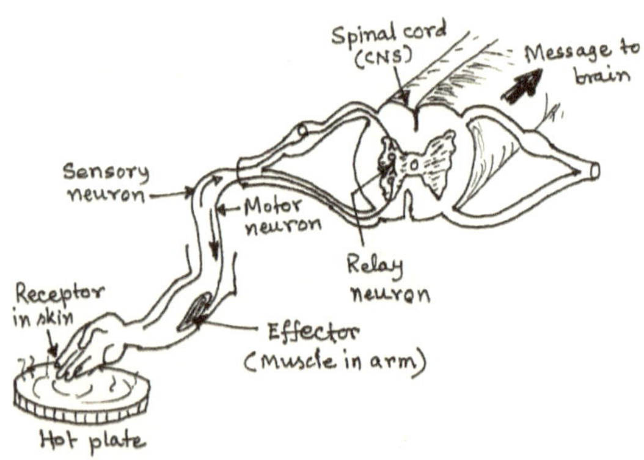

Figure 3.4. A schematic diagram of the reflex pathway the called *reflex arc.*

REFLEX PATHWAYS: In this process, the sensory neurons pick up a signal from a receptor (a sensory organ) and send it to the relay neuron in the spinal cord through the shortest neural path. The spinal cord immediately instructs the muscles (the *effectors*) through the motor neurons. The reflex arc is more direct than the CNS-PNS pathway of the brain and the signal processing time is instant. The muscles attached to the effectors, contract making the hand move away. There are four reflex pathways, created by the cytoplasmic proteins: *respiration pathways, digestion pathways, muscle contraction pathways*, and *Krebs cycle pathways*. The Krebs cycle pathway involves energy-generating processes. In the above example, it is the muscle contraction pathway that processes the input signal.

However, the brain has involvement in versatile activities in association with sensory organs like eye, ear, nose, tongue, and skin, which have been discussed in the subsequent sections.

Vision: eye-brain coordination

When light from an object passes through the cornea of our eyes and falls on the retina, it creates an electrochemical signal. That signal reaches the brain where it is analysed to recognise the object by its colour or shape. If the brain fails to do it, either partial blindness such as colour blindness, or complete blindness arises.

How does the brain recognise the colour of an object? It requires eyes-brain coordination. Sir Isaac Newton discovered that a thing does not have a characteristic colour. It appears coloured if it reflects rays of a specific colour from the incident white light, and absorbs the remaining. An apple appears red because it reflects only the red colour and absorbs other colours of the incident light. An object will appear white if it reflects all the colours of an incident white light and will appear black if it absorbs light completely. It is an optical illusion! In reality, some pigments are responsible for bringing specific colours. Red apples get their colour from a pigment called *anthocyanins*. That pigment develops as the apple grows. It is present in many other fruits and vegetables such as cranberries, raspberries, cherries, cabbage, and other red or purple foods.

Figure 3.5 shows the internal structure of an eye. Retina has a layer of light-sensitive receptors on the inner wall of the eye. Those receptors generate an electrochemical signal, which goes to the specific area in the brain and produces familiar sensations of colour.

Retina has two types of light-sensitive cells (called *photoreceptors*), *rod cells* at the edges, and *cone cells* at the centre. Rod cells are active in low light conditions, while cone cells in bright light conditions. Cone cells have three photo-molecular pigments for red, green, and blue. Those pigments are sensitive, respectively, to long (L), medium (M), and short (S) wavelengths of light, respectively. The photo-molecules are proteins, called *opsins*, such as rhodopsin (for night vision) for rod cells. Photo-molecules have different rates of absorption of light photons; photons of different wavelengths also have different probabilities of getting absorbed. When light falls on the retina, the photo-molecules absorb different colours at different rates and produce corresponding absorption peaks in the opsins with separation as close as 1.0 nm. The human eye is sensitive enough

to distinguish such minute separation of colours.

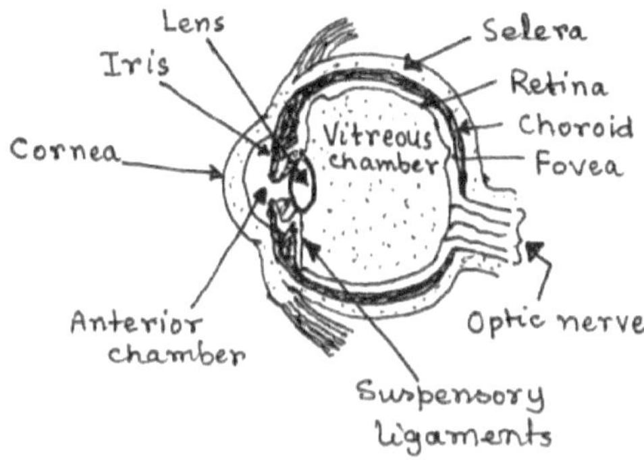

Figure 3.5. The inner structure of the human eye.

Upon absorbing light, opsins undergo an instantaneous reversible chemical change that triggers electrochemical signals of specific codes. Those codes are generated by the retinal ganglions by comparing the signals between neighbouring cones (or rods in low light). For example, light from a red object activates only the red-pigmented L-cones in the retina. The coded electrochemical signal is then transported to the thalamus, just above the brainstem, through the optic nerves (see Figure 3.5), also known as the *cranial nerves II*. Subsequently, the thalamus relays the signal from the lateral geniculate nucleus to the primary visual cortex (called hv4) in the calcarine sulcus of the occipital lobe. The primary visual cortex, which is called *retinoptic organisation*, processes the signal and sends it to the other extrastriate cortex areas (called v4α). There it is further processed to generate the colour perception. Collectively, hv4 and V4α are known as the *v4-complex*, which develops colour images. The visual cortex, as a whole, is known as the *colour centre*. The perception of colour helps to visualise the colour of the object, e.g., a red apple. In the low light condition, the same red colour would stimulate only the rod cells in the retina, which are not colour sensitive, and the object will appear as grey. However, the stored knowledge would influence the perception of colour. Hence, the observer could realise the real colour of the article, for example, a red apple. It is known as *colour constancy*.

COLOUR MATHEMATICS: The primary visual cortex has two types of colour-sensitive neurons: a *single opponent* and a *double opponent*. The single-opponent responds to a broader region of colours in the signal, and the double-opponent responds to the boundaries of different colours. The double-opponent is receptive to complementary inputs from various cones, which is ideal for identifying the contrasting colours such as red-green. It calculates the local cone ratios in the input signal. The single-opponent splits into two categories. The input from M-cones complements the L-M neurons that receive an input signal from the L-cones. The sum of the signals from both L- and M- cones complements S/(L+M) neurons that receive an input signal from the S-cones. Such a complemented combination of colours allows the visual cortex to interpret differences between the colours through a logical calculation in the brain. The brain also recognises other characters of an object, such as size and shape.

Our vision system is very delicate. Any damage anywhere might cause partial such as colour blindness or complete blindness. In the book (2019), *The Neuroscientist who lost her mind: A memoir of madness and recovery*, Barbara K. Lipska has

described the loss of her eyesight, at the lower right quadrant of the vision area, while typing on a computer keyboard. Being a neuroscientist and brain specialist, she suspected a tumour (due to metastasis of her earlier skin melanoma) in the frontal cortex, which was proved correct through the MRI scan.

Hearing: ear-brain coordination

Which part of our brain is responsible for the interpretation of sound? It is the auditory cortex in the temporal lobes, in the limbic systems, which analyses of a sound such as music, voice, or noise.

Our audio receiver ear has three parts, outer ear, middle ear, and inner ear. The inner ear, as illustrated in Figure 3.6 (a), consists of *the cochlea*, the balance mechanism, *vestibular nerves*, and a *cochlear nerve*. The cochlea is responsible for hearing. It is a spiral-shaped cavity in the bony labyrinth, filled with a fluid called *perilymph*. It has two nearby membranes, the Reissner, and the basilar, which form partition walls in the cavity, as illustrated in Figure 3.6 (b). There is a little hole in each barrier wall that allows perilymph to flow freely from one compartment to the other and the vibrations to transmit from the oval window through the perilymph.

HEARING STEPS: When sound enters through the auditory canal in the outer ear and strikes on the tympanic membrane, or the *eardrum*, which vibrates. That quavering causes movements of three ossicles, *malleus*, *stapes*, and *incus*. Due to those synchronised movements, stapes presses into a thin membrane of the cochlea, known as the *oval window*. The vibration of the oval window membrane pushes the perilymph fluid inside the cochlea. As a result, thousands of micron size hair bundles, known as *stereocilia*, execute mechanical motions like fluttering back and forth, in response to the flow of the perilymph. Inside the partition wall of the cochlea, stereocilia arranged in rows on top of the hair cells. The movements of stereocilia create a tension in fine tip-links (made of cadherin-23 and protocadherin-15 proteins) that tether each cilium to its neighbour to pull open the channels. Those unlatched channels allow the flow of positive ions (primarily, K^+ and Ca^{2+}) into the receptor cells or hair cells (Muller, 2008). It generates electrochemical signals, which travel through the auditory nerves and reach the auditory cortex of the brain and get translated into sound. Hair cells act as a mechanotransducer. It converts mechanical trembles into an electrochemical signal. As the message reaches numerous relay stations, the brain decodes them as soft, loud, high pitch, low pitch, or location of the sound source to develop a conscious perception. Other areas in the brain recognise the sound by comparing it with the stored experiences and determine an appropriate voluntary response.

A research team at MIT discovered that sound processing in the auditory cortex goes in different stages, starting with the analyses of its low-level features, e.g., the loudness and the pitch. Subsequently, the high-level features, e.g., the source of the sound and its identity, are analysed step-by-step. Ultimately,

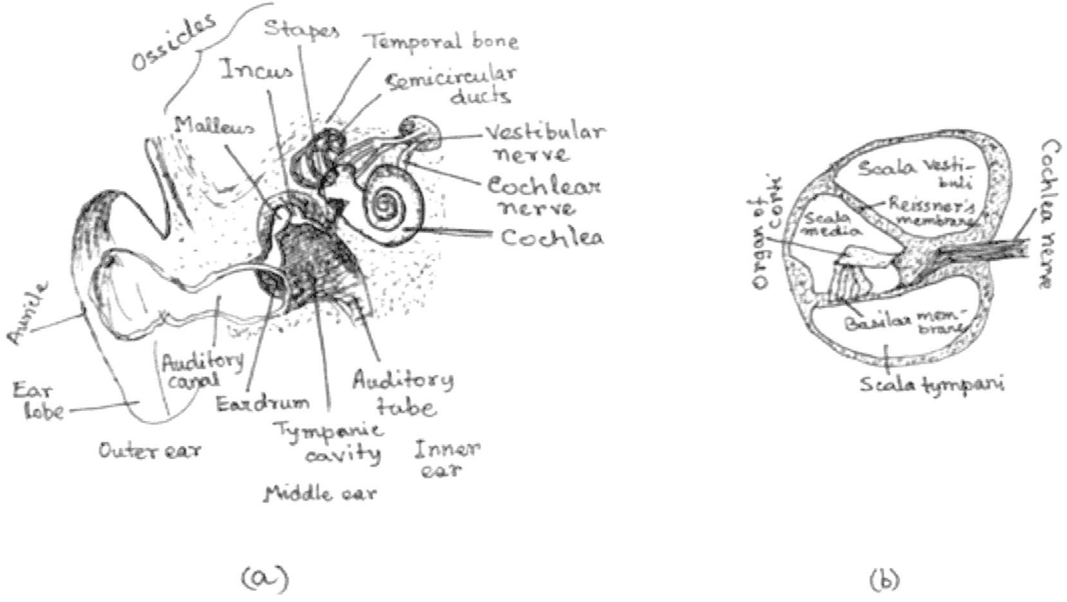

Figure 3.6. Structure of the innermost part of the human ear.

using the stored experience, the sound is recognised. Without past experiences, recognition fails.

EFFECT OF MUSIC IN MIND: Music has a positive result on our minds. Thus, we like it. According to a recent study, listening to music releases dopamine in the brain, which influences our cognitive, emotional, and behavioural functioning. It induces a reward experience (Ferreri et al., 2019). Quoting from a paper by Salimpoor et al. (2011),

"Music is a pattern. As we listen, we're constantly anticipating what melodies, harmonies, and rhythms may come next...."

MUSIC THERAPY: The above quotation explains why we prefer listening to the most familiar music. Dopamine, a neurotransmitter, acts as the bridge between the neurons that make a continuous signalling pathway in the brain. Therefore, music can trigger the flow of chemical signals in the brain. Music-therapy has now accepted an alternative method of treating neurological or psychiatric disease. *Can music transform a "bad" gene to a "good" one?* We do not know. Keep this question open for future genetic researchers. However, there is a new development after the central dogma.

CENTAL DOGMA IS NO MORE! In the late 1960s, geneticist Howard Temin published his research that suggested RNA information could flow backward and alter the host cell's DNA code (Lipton and Bhaerman, 2017). It is called *reverse transcription,* and the enzyme that carries it is called *reverse transcriptase,* which was discovered by Temin. He shared Nobel

Prize in Physiology in 1975 for that work. So, like the environmental effect on DNA (epigenetic), the musical efficacy may also produce DNA mutation. It deserves a detailed investigation.

Smelling: nose-brain coordination

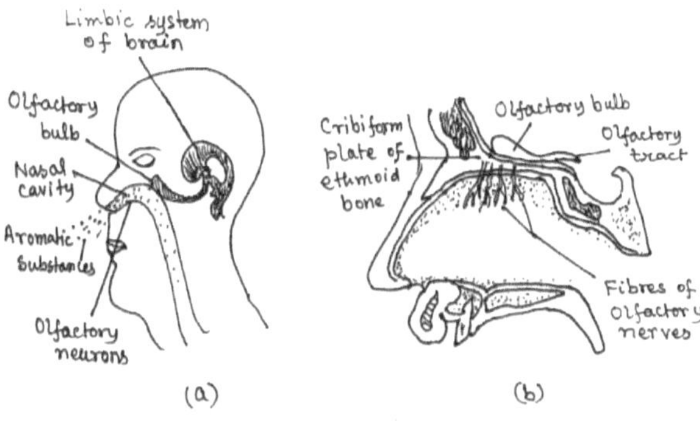

Figure 3.7. The internal structure of the nasal duct and its connection with the brain.

The smell is a chemical sensation in the limbic system of our brain. The organ system that processes and recognises odor is known as the *olfactory system*. A sketch of it has shown in Figure 3.7.

SMELLING STEPS: The processing of smell takes a few steps: first, odor molecules (such as amines, ketones, or esters) reach the nostrils through airflow and get dissolved into its mucus.

Second, a set of specialised receptor cells (known as *olfactory neurons*) in the olfactory epithelium bind the dissolved molecules. More specifically, proteins at the tip of the olfactory nerves, known as the *cranial nerves-I,* bind odor molecules; the binding strength depends on the odor molecule. The complexity of the receptors and their interactions with the odor molecules allow the brain to recognise various smells.

In the third step, the interaction sets an action potential across the sensory neurons. The chemical signal travels along the axon and reaches the glomerulus cells of the olfactory bulb at the back of the nose.

The nsigal further proceeds to the mitral cells of the olfactory bulb, which delivers the signal all over the brain.

In the fifth stage, neurons in the olfactory bulb change timing and frequency of the neuronal spikes of the odor signal, which enable the brain to differentiate between the odors. This process involves several areas in the brain such as amygdala, hippocampus, and orbitofrontal cortex.

After analysing the odor signals, the brain compares them with the perceptual memory (established by the prior exposure) and accordingly generates different sensations of smell.

The smelling sensations may vary from person to person. It depends on several factors, such as the binding of specific odorants to specific receptors. A person has about 400 functional odorant receptors. Then the translation of the odorant-receptor interaction by the brain and psychological and emotional states in mind. For example, a vegetarian dislike the smell of cooked meat, while a non-vegetarian finds it delicious. Dysfunction of any part of the olfactory system, including portions of the brain, causes failing to sense the smell.

Taste: tongue-brain coordination

Is the tongue so pivotal? Yes, we can't talk or enjoy the taste of food without it. You might have seen the iconic tongue-stuck-out-photo of the eminent physicist of the 20th century, Professor Albert Einstein. On 14th March 1951, in his 72nd birthday party that was arranged by the staff at Princeton, Einstein met the United Press photographer, Arthur Sasse, who took several photos of him. At the end of the party, when tired Einstein entered his chauffeured car, Sasse requested his pose for once more. This time Einstein was reluctant, and he turned towards him and stuck out his tongue. That explained his mood!

TOUNG STRUCTURE: Figure 3.8 shows the sectional view of the taste buds in the tongue. The tongue surface, along with the rest of the oral cavity, is lined by a stratified squamous epithelium. The raised bumps, called *papillae* (singular = papilla), on the tongue surface, have the structures for gustatory transduction, located in the borders of papillae. They increase the surface area of the tongue to make sure that individual tastes can be perceived more intensely. Each papilla contains 3–5 taste buds, implanted in the papillae, each of which has sensory cells connected to several sensory nerve fibres. At the tip of each taste bud, there is a taste pore that works as a fluid-filled funnel. This funnel contains many thin hair-like sensory cells, called taste hairs, and the proteins on them bind chemically with the food molecules.

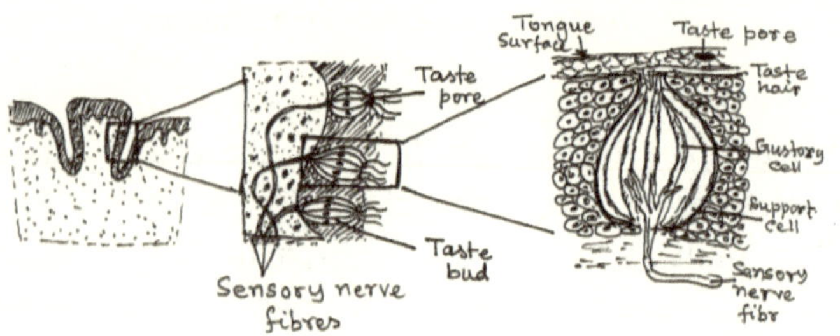

Figure 3.8. The internal structure of the taste bud on the tongue.

TASTING STEPS: The mechanism of sensing taste by the brain is a process of tongue-brain coordination. The perception of taste is engendered by the *gustatory system* when a food substance in the mouth interacts chemically with the taste receptor cells in the taste buds in our tongue. The chemical signal arising due to the binding interactions, first, gets transferred to the cranial nerves and then to the medulla oblongata (at the lower section of the brainstem) through the sensory nerve fibres attached to the taste hairs. Afterward, the signal flows to different parts of the brain concerned with the sensory perception, where it is combined with smell signals to generate a delicate feeling about the foodstuffs.

Edibles could be of different tastes (depending on their ingredients), which have five general qualities: salty, sour, bitter, sweet, and umami. Umami is the Japanese term for savoury sensation. There are some significant reasons behind the taste sensation. For example, salty and sour detection helps to control the *salt and acid balance* in our body. Bitter taste warns us of poisons in the foods. Sweet taste provides a guide to the intake of calorie-rich foods, and umami (the taste of the amino acid glutamate) may flag up protein-rich foods. Besides, without the savour, cuisines

would appear repetitive for eating. The recognition of taste has been nicely described in a paper by Bernd Lindemann (2001). He explained the simple goal of our sense of taste as,

"Food is already in the mouth. We just have to decide whether to swallow or spit it out. It's an extremely important decision, but it can be made based on a few taste qualities".

Skin: the largest sensory organ

Skin is the largest and vital organ in our body. It regulates the body temperature, prevents dehydration, protects from external heat, cold, and UV radiation. It is also a storeroom of our body water, fat, and metabolic products. Upon exposure to the sunlight, healthy skin produces vitamin D that is essential for bone development and other functions.

Figure 3.9 shows the internal structure of the skin. It has three layers of tissues, namely, *epidermis* (outer), *dermis* (middle), and *hypodermis*, or *subcutaneous* (inner). The epidermis layer contains no blood vessel but small nerve cell endings, which are in constant contact with the brain. It also consists of sub-layers. Pigments (e.g., melanin) are produced at the inner region of the epidermis to accord a colour to the skin. Dark and brown skins have more melanin. It also protects the epidermis from the UV rays. Keratin proteins get produced to strengthen the skin.

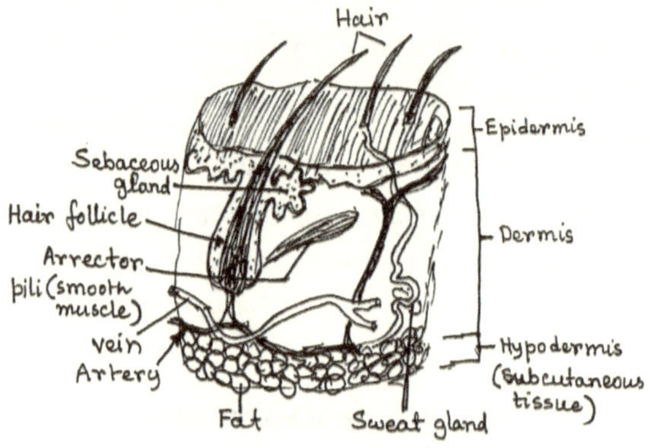

Figure 3.9. The internal structure of human skin.

The dermis layer is made of collagen and elastin proteins to ensure strength, flexibility, and morphology in the skin. It contains blood and lymph vessels, fat, oil (sebaceous) and sweat glands, nerve fibers, sensory receptors, and hair follicles. Sebaceous glands are usually attached to hair follicles and release a fatty substance (sebum) into the follicular duct and thence to the surface of the skin. It helps to soften and lubricate the skin and the hair. Hairs help to lubricate the scalp that secretes pheromones and cools or warms our heads. Arrector pili muscles generate heat when the body becomes cold, and contract to straighten hairs up on the skin. The nerve endings (corpuscles) in this layer provide the sensations of touch, pain, temperature, and pressure. The hypodermis layer contains a fat layer that acts as the shock absorber to protect vital organs in the body and the storeroom of energy.

SKIN DISCLOSES MENTAL STRESS: Skin is a sensitive perceiver, as well as the reflector of our psychological stress in mind. Psychological stress arises when people are under mental, physical, or emotional pressure (Chen et al., 2014). The brain perceives it and, accordingly, releases stress hormones (e.g., cortisol), which triggers a wide range of physiological and behavioural changes and also the responses that help our body to adapt the stress (Papadimitriou et al., 2009). The stress perception passes from the brain to the skin through the hypothalamic-pituitary-adrenal (HPA) axis. It is the pathway of direct influences and feedback interactions between the three components: hypothalamus, pituitary gland, and adrenal glands. The pituitary gland has a location below the hypothalamus and the adrenal glands on top of the kidneys.

HPA ACTION STEPS: First, the hypothalamus releases the corticotrophin-releasing factor (CRF). Then CRF triggers the release of adrenocorticotropic hormone (ACTH) from the pituitary gland. Eventually, the adrenal cortex releases glucocorticoids that produce an array of upshots. Sequels of that include mobilizing energy into the bloodstream from the storage sites in the body, increasing the cardiovascular tone, and delaying the long-term processes in the body that are not essential during a crisis, e.g., feeding and digestion.

STRESS INFLAMMATIONS: Under stress, the skin may turn pale or lighter due to the rushing of blood away from the skin to the heart causing it to lose some natural pigments. Stress may also cause the face to turn red or pink by dilating the blood capillaries (due to electrostatic force in the muscle fibers) inside the facial skin, which allows more blood to reach in the skin. Stress may cause skin problems or skin disorders. Overstress may move the blood away, causing insufficient oxygen supply and reducing the nourishment in the skin. Thus, the skin becomes pale and dull. Stress and anxiety may also cause liver dysfunction, which may lead to jaundice and, in effect, yellowing of the skin.

A human body has many organ systems. I have discussed, so far, about the nervous system and all the associated sensory organs. All organs do interact with each other and with the brain using the electrochemical signals. There are eleven other major organ systems in our body, which are functionally interconnected. I shall discuss those organ systems as per the theme of this book. Note that the skin, nails, and hairs are parts of the *integumentary system*, the functions of which is to act as a barrier to protect the body from external threats, such as viral attack and pollution. It also helps to retain body fluids, protect against diseases, eliminate waste products, and regulate body temperature.

How the body did acquire such well-engineered organ systems? The simple answer is through continuous grappling against the environmental hardships. It proves once again, the phrase, *survival of the fittest* that originated from the Darwinian evolutionary theory. Famous English philosopher, biologist, anthropologist, and sociologist, Herbert Spencer remarked in his book, *Principles of Biology* (1864),

"This survival of the fittest, which I have here sought to express in mechanical terms, is that which Mr. Darwin has called 'natural selection', or the preservation of favoured races in the struggle for life"

Skeleton system

The skeleton system gives our body a fixed structure and outlook. It also performs five essential functions: body movements, protection of internal organisms such as the heart, lungs, and the brain, production of blood cells, storage of minerals (e.g., calcium), and endocrine regulation. An adult skeletal system contains 206 bones; those bones are joined through the ligaments and supported by the muscles. Ligaments are fibrous collagenous connective soft tissues between the bones, which usually serve to hold the structure together and maintain stability. The bone-muscle anatomy has formed by another collagenous connective fibrous soft tissues, called a tendon.

Collagen is a fibrous protein with high tensile strength. When we move our limbs, ligaments absorb the stress between the two adjacent bones and, thus, maintain the structure at the joint. In other words, it serves as a cushioning pad for the adjacent bones.

OLD AGE ISSUE: At old age, the ligament at the joint loses its elastic property (i.e., flexibility) or wears away and becomes weak in providing necessary structural support. Consequently, there will be friction between adjoining bones and inflammation. It causes joint pain, which is known as *arthralgia*.

What is the force that makes muscles work? It is the electrostatic force that causes attraction or repulsions between the muscle proteins. Such interactions produce elastic muscle movement. With aging, proteins get denatured and lose their surface charges. Consequently, intra-muscle electrostatic interactions become weak, and joint ligaments lose their elastic property. That causes friction between the cup-bones giving rise to joint pains. There could be another reason for causing joint pain. Ligament injury or infection may also cause other symptoms.

RELIEF MECHANISM: The nonsteroidal anti-inflammatory drugs, such as ibuprofen and diclofenac, can control the pain by prohibiting *cyclooxygenases* (COX) enzyme from producing prostaglandins, which promote inflammation like pain or fever. In consequence, symptoms of inflammation reduce. But at the same time, those drugs have many side effects in the stomach and intestines.

INNER STRUCTURE OF BONE: Bones are made of living cells and inorganic salts of hydroxyapatite, calcium, and phosphate, and embedded in a mineralised organic matrix, primarily composed of collagen. Bone has a pile of three layers: *compact-bone* covers the outer part, *spongy-bone* in the middle, and *bone marrow* in the core, as shown in Figure 3.10.

The compact-bone, which forms eighty percent of the structure, protects the inner part that performs vital functions like bone marrow production. The compact-bone primarily consists of cells, called *osteocytes*. The gaps between the cells create microscopic channels for nerves and blood vessels to pass through (see Figure 3.10).

The remaining twenty percent of spongy-bone contains bone marrow, nerves, and blood vessels. There are two types of bone marrow, red and yellow in half. The red bone marrow is also called myeloid tissue. It constitutes flat bones such as shoulder blades and ribs, where red blood cells (RBC or erythrocytes), white blood cells (WBC or leukocytes), and platelets get produced. RBC is red due to the presence of iron-rich protein, called *haemoglobin*, in it.

The yellow bone marrow is also called fatty tissue. It composes long bones such as thigh bones and contains a low level of haemoglobin. It also consists of a high level of carotenoids that give its yellow colour.

Blood vessels run through both types of bone marrows to deliver nutrients and to remove waste from the bones. Other than osteocytes, bone contains another three types of cells: osteoblasts, osteoclasts, and lining cells. Osteoblasts create new or repair existing bones. Osteoclast cells break down existing bone material and reabsorb it. It also works for healing and reshaping bone after a break, together with osteoblasts. Lining cells cover the outer surface of the compact-bones and control the movements of minerals, a clan of cellular and other materials into and out of the bones.

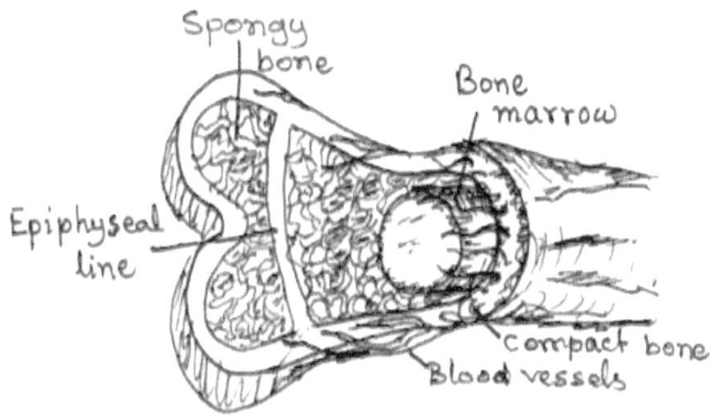

Figure 3.10. The internal structure of bone.

STEM CELLS: The matured cells have specific functions in the body. But stem cells[7], being immature, do not have any physiological activity. They are malleable to transform into any other cell as per the requirement. Stem cells are also available in the brain, in skeletal muscles, in the liver, in the skin, and some other tissues. Bone marrow comprises two different stem cells, *mesenchymal* and *hematopoietic*. The red bone marrow consists of hematopoietic stem cells that produce blood cells and platelets. When a stem cell divides, at first, it becomes an immature cell, which again divides and matures into either red blood cell, white blood cell, or platelet. The yellow bone marrow contains mesenchymal stem cells which produce fat, cartilage, and bone.

MUSCULAR SYSTEM: It contains smooth and cardiac muscles. Muscles are soft tissues consisting of protein filaments of actin and myosin. Those proteins are flexible and contract to change the length and shape of the cell. Thus, muscles work to induce forces, locomotion, and movement of internal organs. Muscle functions include contraction and expansion of the heart and lungs, movement of food through the digestive system via peristalsis, and many others. A human body is all-compassing with several types of muscles that help to build strength in the body, storing energy, doing hard work, and giving a muscular outlook. Like most organ systems, the muscular system also does not work alone, but in collaboration with the other organ systems such as the skeletal system, nervous system, circulatory system, digestive system, respiratory system, and immune system.

Circulatory system

The circulatory system is a cardinal organ system that carries blood between the heart and body cells through closed paths of arteries and veins. The heart is the driver of this organ

system and works as a two-way pump. This process is called *perfusion*. Figure 3.11(a) shows the blood circulatory pathways from the heart to various organisms and back, and (b) the internal structure of the heart.

COMPOSITION AND FUNCTIONS OF BLOOD: Blood is a colloidal fluid. It is composed of fifty-five percent of plasma (a light-yellow liquid) and forty-five percent of blood cells. Blood carries some *essential items* to deliver to body cells in exchange for some waste products, which are like *garbage*. Thus, the blood performs delivery and pick up services for cells. Essential items include oxygen, vitamins, nutrients, and hormones.

While passing by the lungs, blood collects oxygen (O_2) from the air in alveolar sacs in the lungs. The lungs fill the sacs with O_2-rich air while inhaling. This way, the circulatory system collaborates with the respiratory system using blood as the mediator. The alveolar sacs have thin walls wrapped with capillaries of arteries and veins.

As fresh blood reaches the lungs, O_2 diffuses spontaneously from the air in the sacs and then into the blood through the capillary walls. The RBCs in blood consist of haemoglobin (Hb), which has four iron (Fe) atoms that bind four oxygen molecules by covalent bonds. After accumulating O_2, blood comes to the left atrium of the heart through the pulmonary vein, and then to the left ventricle, passing through the mitral valve. From the left ventricle, O_2-rich blood crosses the aortic valve and goes to body cells through the arteries. Eventually, O_2 molecules dissociate from RBCs, diffuse out of the blood capillaries and enter into the body cells through their plasma membranes. If the blood lacks sufficient Hb or Fe, body cells will suffer from O_2 deficiency for the metabolism.

Glucose and amino acids, require for metabolism, come from the foods and beverages. In our digestive system, foodstuffs break down into those nutrients that get absorbed by the blood plasma at the small intestine. The thin walls of the mucosa, in the gut, are wrapped with blood capillaries. Nutrient molecules pass through those walls and get dissolved in the blood plasma through passive diffusion, which again helps nutrients to enter into the body cells from the blood capillaries. At the plant roots, the diffusion process helps absorbing salts, and the osmosis process helps to suck up water from the soil.

Diffusion and osmosis are two contrary processes. In the diffusion process, solute molecules move from higher concentration regions to lower concentration regions in the solutions. Conversely, in the osmosis process, the solvent molecules in the solution move from low to high concentration regions through a semi-permeable membrane. A passive diffusion or osmosis process requires no, but an active process needs energy exchange.

In the cell, mitochondria produce ATP from ADP through a reaction between glucose ($C_6H_{12}O_6$) and O_2 during *respiration*. It is part of the *metabolic process*. ATP or adenosine triphosphate is an energy molecule, which gets synthesized in bulk by oxidative phosphorylation.

Metabolism is a combined process of *catabolism* and anabolism. Catabolism is the process of generating energy from food. Anabolism is the process of repairing cells or building new cells using food nutrients (mainly amino acids). In the anabolic process, carbon dioxide (CO_2) and water (H_2O) get produced as the by-products. CO_2 is toxic for cells and, thus, considered as

a waste product, which the cells discard as garbage. The blood collects CO_2 from the cells and propels into the lungs. Lungs throw out CO_2 to the atmosphere during exhalation.

CO_2 gets transported in three ways: a small number by dissolving in the blood plasma, and by binding with the haemoglobin in RBCs, while the remaining large number as bicarbonate in the RBCs. In RBC (i.e., erythrocytes), carbonic anhydrase catalyses the reaction $H_2O + CO_2 \rightarrow H_2CO_3$. In an aqueous medium, H_2CO_3 dissociates into H^+ and HCO_3^- (bicarbonate) ions. HCO_3^- ions diffuse out from RBCs to the blood plasma in exchange for chloride ions (Cl^-) that diffuse in the opposite direction. This process is called the *chloride shift*. In RBCs, H^+ ions react with O_2-bound haemoglobin ($Hb.O_2$) and release out O_2 following the reaction: $4Hb.O_2 + H^+ \rightarrow 4H.Hb^+ + 4O_2$, and reduce the acidity in the blood.

Other waste materials such as urea (in the liver) and lactic acid (in the muscles and many other cells), produced during the metabolic activities, get collected by the blood and dumped at respective organs; for example, urea at the kidney. Thus, RBCs actively participate in metabolism.

WBCs are the stakeholders of the immune system that is responsible for the defence activities in the body against infectious pathogens and foreign invaders. Platelets react to bleeding from blood vessel injury by clumping, thereby initiating a blood clot.

LIFE CYCLES OF BLOOD CELLS AND PLATELETS: The life of the blood cells and platelet are short; RBC has around 120 days of life, WBC has few hours to a couple of days, and

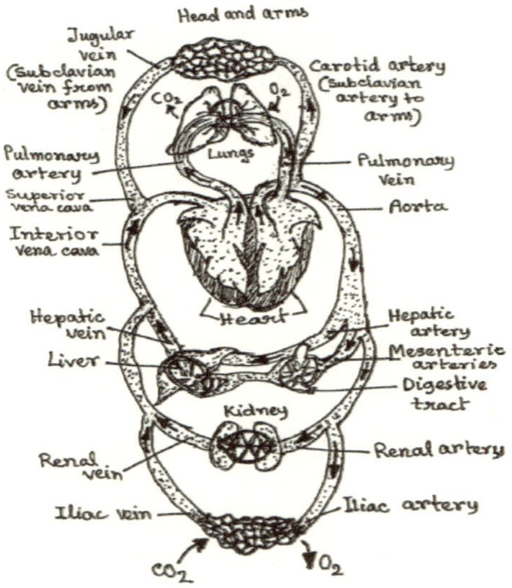

Figure 3.11. (a) Circulation pathways of blood from the heart to various organisms and back, (b) internal structure of heart showing SAN, AVN and Purkinje fibres.

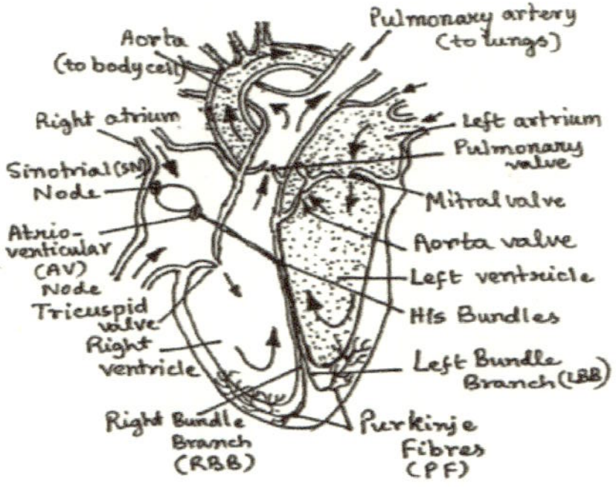

platelet around ten days. Therefore, new RBCs, WBCs, and platelets get manufactured in bone marrows. The rate of production of the blood cells depends on the body's need. For example, when the O_2 content in the tissues becomes low, or the number of RBCs in the blood decreases, the kidneys generate and release a hormone called erythropoietin. It stimulates the bone marrows to produce more RBCs. Similarly, the bone marrow manufactures and releases more WBCs in response to infections, or platelets are made and released in the blood in response to bleeding.

DISEASES AND MEDICINE: The iron deficiency in the blood usually causes RBC disorder, as iron is the main component of hemoglobin. This situation causes *anemia*. In such cases, doctors may prescribe medicines/diets for the iron supplement to the patient. In severe iron-deficiency, the patient may require intravenous iron injection or a blood transfusion. The over or underproduction of WBCs cause several diseases. Leukocytosis occurs if the number of WBCs increases in the blood. Autoimmune neutropenia arises due to a low count of the bacteria-fighting neutrophil cell. The low count of platelet or platelet function disorder causes blood clotting problems, i.e., an increased tendency to bleed or bruise; for example, *hemophilia*.

The pumping of the heart, to circulate blood throughout the body, is called the heartbeat. It makes a "lub dub" sound. The heartbeat rate is measured as beats per minute or bpm. For an adult, the heartbeat usually varies in the range of 70–75 bpm, in rest condition.

How does the heart pump spontaneously? It is due to the isovolumetric contraction of the ventricular muscles, producing the "lub" sound. Contraction is produced by a spontaneous electrical impulse from some clusters of cells. The heart muscles relax again, making the "dub" sound when the pressure in the ventricles is less than arteries. There are many clusters of cells. One is called the *sinoatrial* (SA) *node* or *sinus node,* which belongs to the upper part of the wall of the right atrium of the heart (see Figure 3.11b). The electrical impulses in the SA node, get originated from the sequential activation of various ion-currents in its constituent cells. Those cells contain Ca^{2+} ions, and the heartbeat rate gets influenced by the transient changes of those ions. Those impulses conduct through few channels from the SA node to the other node, called the *atrioventricular* (AV) *node.* The AV node serves as an electrical relay station, slowing down the electrical current sent by the SA node before passing it to the ventricular muscle cells. From the AV node, a tract of conducting fibres, called *atrioventricular bundle* or the *bundle of His,* runs through the cardiac muscle to the top of the ventricular septum that separates the two ventricles of the heart. Then it branches out to form the right and the left bundle branches. Each branch finally divides into subbranches, called *Purkinje's fibres* (PFs). Those capillary fibres pass into the cells of the ventricles. The electrical impulses generated in the SA node pass-through that cardiac network as an action potential and activates the ventricular muscle cells to contract and push the blood out of the heart. When those cardiac muscles relax, the deoxygenated blood, coming from the body cells, enters into the right atrium and then into the right ventricle through the tricuspid valve.

During contraction, the deoxygenated (CO_2 rich) blood gets pushed through the pulmonary valve to the pulmonary artery, and finally to the lungs to release CO_2. The lungs exhale CO_2 to the

atmosphere by squeezing sacs through muscle contraction. Thus, the cardiac network of SA node, AV node, and Purkinje's fibre is called the *natural pacemaker*. Several physiological functions such as metabolism, autoimmune, and detoxification that are mediated by the bloodstream, get controlled through the heartbeat due to spontaneously generating and passing the activation potential or impulse through the cells of the natural pacemaker network. Thus, if the heart is in good condition, the immune system will work to protect the body from diseases. The neural signalling process also gets mediated by the generation and passage of the activation potential through the neural network. Whether cardiac or neural, signals are nothing but electrochemical currents of ions through specific cells.

Endocrine system

The endocrine system is a network of glands. It produces hormones to regulate multiple physiological functions such as metabolism, tissue functions that are related to growth, development, and reproduction, as well as mental states. Hormones, secreted from the glands and carry specific instruction, are directly released into the extracellular environment from where they diffuse into the bloodstream through the blood capillary walls. Through the bloodstream, hormones get transported to the target sites.

There are two types of hormones, polar and non-polar. The non-polar hormone (NH) molecules, such as steroids, are soluble in lipid. Thus, NH can diffuse through the lipid bilayer membrane into the cytoplasm of a cell. Either in the cytoplasm or nucleus, the NH molecules bind with the matching receptors for specific action as per the instruction associated with them. The polar hormone (PH) molecules such as amino acid derivatives, on the other hand, cannot pass through the cell membrane on their own. So, they bind with the specific receptor proteins in the targeted cell membrane. Consequently, some conformational changes appear in the inner parts of the receptors that activate a cascade of enzyme activity within the cell. The cell, then, acts as per instructions given in the PH.

Figure 3.12 shows the (a) endocrine system of glands and (b) schematic transportation pathways of hormones from the secreting glands to the target cells through the bloodstream. A human body has five types of secreting glands: (i) *hypothalamus, pituitary,* and *pineal* in the brain, (ii) *thyroid* and *parathyroid* in the neck, (iii) *thymus* in the lungs (iv) *adrenals* on top of the kidneys, and (v) *pancreas* behind the stomach and *ovaries* or *testes* in the pelvic region.

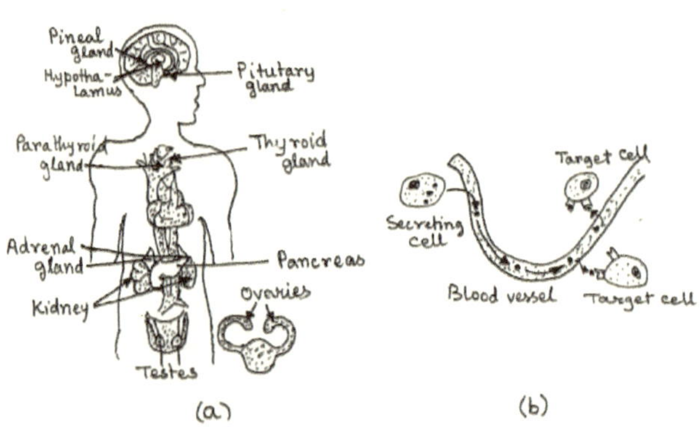

Figure 3.12. (a) Endocrine glands at different locations, and (b) transportation of hormones from the secreting gland cells to the target cells through the bloodstream, where they bind with the cell membrane receptors.

HORMONES AND THEIR FUNCTIONS: Hypothalamus is a connecting organ between the endocrine system and the nervous system, located at the lower central part of the brain. It secretes hormones with necessary instructions in its structure to start and stop secretion from the pituitary gland that has located beneath the hypothalamus. The pituitary hormones are growth hormones (GH) that motivates other organs to carry out their specific activities. For example, prolactin prompts mammary glands in the breast to produce milk. Likewise, luteinizing (LH) and follicle-stimulating (FSH) hormones jointly activate the secretion of oestrogen in women and testosterone in men for sexual development.

The pineal gland, in the middle of the brain, secretes melatonin hormone that controls the *wake-sleep cycle* in the body. On arrival of daylight, retinas in the eyes become active and send a chemical impulse to the suprachiasmatic nucleus (SCN) in the hypothalamus. The SCN then sends signals to trigger other parts of the brain to start their functions, i.e., to secrete hormones to make the body active and raise the body temperature. The SCN simultaneously sends a signal to the pineal gland to stop melatonin secretion until the next darkness appears. At night, the SCN becomes less active, and its inhibitory signal gets reduced. That causes the production and release of melatonin into the bloodstream and the cerebrospinal fluid. Through the blood circulation, melatonin reaches everywhere in the body and binds with the specific receptors at the body cells. Thus, the tissues could detect the peak in the melatonin circulation at night and signal the body to relax. This day-night cycle is called the *circadian rhythm*, and the SCN is known as the "biological clock" as it controls the melatonin secretion.

People who suffer due to insomnia (sleeping disorder) have a reduced level of melatonin secretion. Injection of melatonin in the bloodstream will be a possible remedy for that. Melatonin has many other physiological functions. It is an antioxidant agent. It also influences the immune and reproductive systems.

The thyroid gland, at the lower front part of the neck, has two lobes connected by a strip of tissues, called the isthmus. This gland secretes two thyroid hormones, T3 and T4, which control many vital functions such as the heartbeat rate and the body weight. If there is insufficient production of the thyroid hormones, a condition arises that is called *hypothyroidism*. In that case, many metabolic functions, including the heartbeat, become slow. On the other hand, if there is excess production of thyroid hormones, there will be a situation called *hyperthyroidism* when metabolism, including the heartbeat rate, increases abnormally. The hypothalamus and the pituitary glands interact with the thyroid gland to maintain a balance between T3 and T4. For a reduction (or elevation) in the level of T3 and T4, the pituitary gland gets triggered by the hypothalamus. That releases the thyroid-stimulating hormone (TSH). TSH helps to increase (or decrease) the T3 and T4 secretion from the thyroid gland. The hypothalamus releases the TSH releasing hormone (TRH) for signalling to the pituitary gland. The thyroid gland also produces calcitonin, which reduces the calcium ion (Ca^{2+}) level in the blood plasma in specific situations. There are two pairs of small parathyroid glands, one pair on each lobe of the thyroid gland. These glands produce and secrete the parathyroid hormone for increasing the Ca^{2+} level in the blood plasma if it goes below the minimum level (8.6–10.3 mg/dL).

The thymus gland is located behind the sternum and between the lungs. Its function is to secrete thymosin hormone that stimulates the development of WBCs, known as T-lymphocytes. Those T-cells do fight against the infections. It plays a crucial role in developing a child's immunity. The thymus gland does not work through the entire life, and it starts shrinking after puberty.

There are two triangular adrenal glands on top of each kidney. Each gland has an inner part called the *adrenal medulla*, and the outer part is called the *adrenal cortex*. The adrenal medulla produces the adrenaline hormones to help the body to cope with the physical and emotional stresses by increasing the heartbeat and blood pressure. The adrenal cortex produces the corticosteroid hormones to regulate the metabolism, balance of salt and water in the body, the immune, and the sexual functions. So, the endocrine system does interact with the immune system, the digestive system, and the reproductive system.

Corticosteroids overturn many discomforts, such as allergies, asthma, and atopic dermatitis. There are different corticosteroids like cortisone or hydrocortisone.

How do steroids work in the body? The corticosteroid molecules on reaching the targeted cells in the immune system, bind with the membrane receptors. That reduces their activity of releasing the inflammatory signals. As a result, the symptom of the inflammation as asthma goes off, and the body feels comfortable.

How does the body respond to a sudden threat? In an unforeseen risk, our body suddenly becomes cold and stiff, the heartbeat or the breathing rate increases, and the mental focus sharpens. In response to possible danger, adrenal glands secrete adrenaline that brings the body in a *flight-or-fight* mode. The blood rushes from the hands and toes to the vital organs such as the heart, kidney, and lungs, to increase their energy for fast and efficient functioning. It causes cold feelings at hands, toes, face, and through the spinal cord. It happens in the reflex action. As the body temperature reduces, our brain sends a signal through the neural network to the muscles to contract and relax to restore the standard body temperature. That triggers a feeling of shivering down the spinal cord.

The pancreas is an elongated organ at the back of the abdomen behind the stomach. One part of it, called the *exocrine pancreas*, does secrete the digestive enzymes and the other, called the *endocrine pancreas*, secretes insulin and glucagon. Those regulate the glucose level in the blood. If the pancreas stops producing insulin (a condition of *type-1 diabetes*), the sugar level can get dangerously high. If the pancreas produces insufficient insulin that could not control the blood glucose level, *type-2 diabetes* does appear.

Ovaries or testes are also a part of the endocrine system. For women, ovaries produce *oestrogen* and *progesterone* that help to grow the breasts bigger at puberty, regulating the menstrual cycle, and support a pregnancy. During adolescence of men, testes produce *testosterone* that helps growing facial and body hair, helps in enlarging the penis, and in making sperms.

Immune system

The immune system is the defence system that protects the body from diseases causing invaded infectious elements. Unlike other systems, the immune system does not have its own identity. It consists of the parts of other organ systems such as WBCs, antibodies, complement system, lymphatic system, spleen, thymus, and bone marrow. It is also a part of the lymphatic system.

The lymph nodes[8] (made of lymphoid tissues) create an environment for antigens to interact with the lymphocytes through specific antibodies that induce an immune response in the host organism. It causes symptoms like fever, swelling, or running nose that will trigger our immune system to take necessary actions. Once the pathogen gets bound with the lymphocyte (known as *phagocyte*), it gets engulfed and killed. On the lymphocyte cell membrane, antibodies get produced by the B-cells.

The antibody-antigen binding is nothing but a chemical interaction. There are many such chemical interactions in which the immune system takes part to maintain the defence alert and the rescue mechanism in the body. The skin is the primary defence organ for the body as it directly gets exposed to external threats such as bacteria or viruses.

Can the immune system fail? Acquired immune deficiency syndrome or AIDS is a disorder that reduces the efficiency of the immune system, for example, by killing the T-cells. T-cells are like security personnel in the immune system. The immune system may also have a weak level of defense control due to some genetic history. A reaction may cause allergic inflammation or hay fever because an individual cannot tolerate certain elements, known as allergens. A healthy body can always destroy allergens. One can have an allergy to certain foods, pollens, plants, dust, sunlight, or cold. Sometimes allergic reactions could even be fatal.

Lymphatic system

The lymphatic system is part of the circulatory system, as well as the immune system. It consists of a network of lymphatic vessels throughout the body that carries fluid, called *lymph*. It is like the circulatory system of blood. The word *lymph* has originated from the Latin word *lympha*, which means water.

The lymphatic system has two main parts: primary or central lymphoid organs (CLO) and secondary or peripheral lymphoid organs (PLO). The CLO, which consists of thymus and bone marrow, produce lymphocytes from the immature progenitor cells. Like stem cells, the progenitor cells are also undeveloped though they are more specific than stem cells.

Lymphocytes are the subtype of the white blood cells (WBCs), which consist of both T-cells and B-cells, which are produced in the bone marrow, where only B-cells become matured. T-cells obtain maturity in the thymus gland. Thus, after production in the bone marrow, T-cells are shipped to the thymus. Once becoming matured, both B-cells and T-cells enter into the circulatory system and reach to the PLO in search of pathogens. A pathogen is an organism such as a bacterium, virus, or other microorganisms, or any poison that can cause disease.

The PLO consists of *spleen* and *lymph nodes*. Lymph nodes, where the lymph gets filtered, are the members of a network by the lymphatic vessels. Our body has around five to six hundred lymph

nodes, with some clusters at the neck, armpits, groin, around the gut, and between the legs. The afferent (incoming) lymph vessels bring in lymph with a collection of pathogens, which percolates through the lymph nodes, and passes out through efferent (outgoing) lymph vessels; that's how the lymph gets filtered. After filtration, the lymph enters into the bloodstream along with its constituent proteins.

Respiratory system

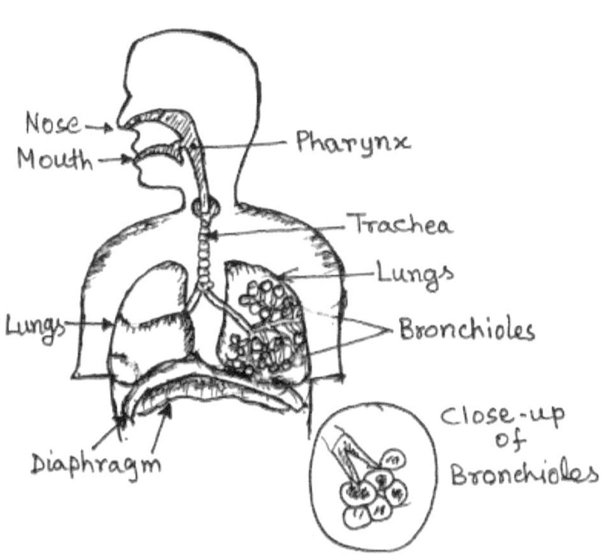

Figure 3.13. Human respiratory system

The respiratory system consists of a series of organs that are responsible for inhaling oxygen (O_2) and exhaling carbon dioxide (CO_2) through the lungs. As mentioned earlier, CO_2 gets produced in the body cells during respiration, which is a toxic element.

Figure 3.13 shows a sketch of the respiratory system. It has three major segments: an airway that includes the nose, mouth, pharynx, larynx, trachea, bronchi and bronchioles, and a pair of lungs and associated muscles. In each lung, bronchi divide into narrower branches and finally into wispy bronchioles. Bronchioles connect to many microscopic air sacs, called *alveoli*, wrapped with blood capillaries.

Air gets inhaled through the nose and the mouth. It contains oxygen. In the respiratory system, air travels down the trachea, i.e., windpipe. Somewhere down the line, the windpipe splits into two stems of bronchi; one goes to the left lung and the other to the right lung where air fills the alveoli. When the blood reaches those alveoli through capillaries, it collects the oxygen that binds with the haemoglobin. Then the blood arrives at the body cells to deliver O_2 for catabolism of food particles. In the catabolic process, CO_2 gets produced as a by-product. While returning, the blood collects CO_2 and dumps it at the lungs. The CO_2-rich air gets exhaled in the atmosphere.

During the inhalation, the muscles between rib cages, the diaphragm, the abdominal muscles, and sometimes the neck muscles tighten and flatten. As a result, the lungs expand and suck the air. When both diaphragm and muscles relax and return to their original shapes, the lungs get compressed and, as a consequence, exhale the degraded air.

Digestive system

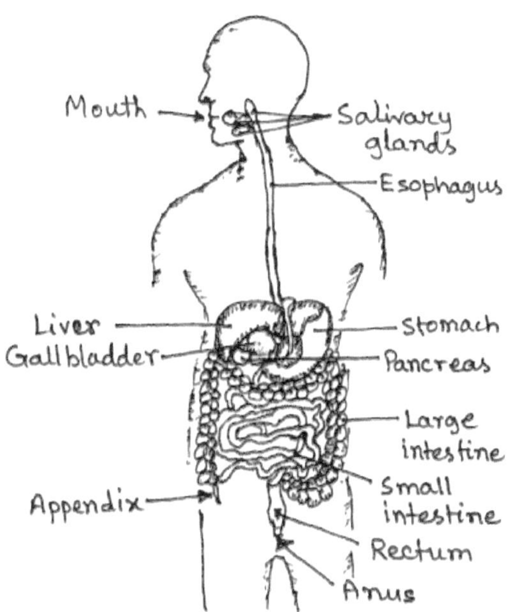

Figure 3.14. Human digestive system.

We take food or health drink, which fulfill vitamins and minerals requirements in the body, energy production, cell growth, and repair, as well as strengthen the immunity system. In the process of digestion, the protein contents in the food break into amino acids, the fat breaks into fatty acid and glycerol, and the carbohydrates break into glucose. For our daily physical works also, we need energy. Energy comes from the oxidation of glucose that occurs during the respiration process in the cells. Besides, waste items are also produced that get excreted from the body.

COMPONENTS OF THE DIGESTIVE SYSTEM: Figure 3.14 shows a sketch of the human digestive system that starts from the mouth and ends at the anus. It has the following segments: gastrointestinal (GI) or digestive tract, liver, pancreas, and gallbladder. The GI tract consists of mouth, tongue, salivary glands, throat or pharynx, oesophagus (i.e., food pipe), stomach, small intestine, large intestine, and anus. The stomach is a food bowel, which has several layers of muscles in its wall. The small intestine is a pipe-like organ of about twenty feet long and an inch in diameter; it has three parts: duodenum, jejunum, and ileum. From the appendix, the large intestine subsequently has cecum, ascending colon, hepatic flexure with a transverse colon, splenic flexure, descending colon, sigmoid colon, and rectum.

PROCESS OF DIGESTION: We chew foodstuffs in the mouth using the teeth. In this process, saliva from the salivary glands comes out to soften and break down the chewed food-grains into small pieces. The amylase enzyme in saliva helps in breaking down the starch molecules into maltose and dextrin. The tongue helps in turning the food upside down to mix with the saliva and make a soft paste that can easily pass through the pharynx region (i.e., throat) to enter into the oesophagus. The oesophagus is a muscular tube that ends at the stomach. As soon as processed food enters the oesophagus, the brain sends a signal to its muscles to contract and relax alternatively along the food pipe to push down the food particles into the stomach. This process is called *peristalsis*.

At the lower end of the oesophagus, a ring-like muscle, called the sphincter, relaxes to pass the food into the stomach. The sphincter usually remains closed to stop the stomach contents

flowing back into the oesophagus like a valve. The stomach secretes hydrochloric acid (HCl) and pepsinogen, an inactive gastric enzyme. Later, it is activated as pepsin by the stomach-acid. Foodstuffs mix with the pepsin and the muscles help to churn it for breaking proteins into fragments of peptides or polypeptides. The process is as follows.

The stomach acid unfolds the protein components of the food, which is called denaturation. Consequently, protein bonds expose, and pepsin takes that opportunity to break them to fragment proteins. Afterward, the stomach pushes the cud into the small intestine. In the duodenum, three main enzymes, pepsin, trypsin, and chymotrypsin (from the pancreas), participate in breaking down the remaining proteins into peptides and polypeptides, which eventually break into amino acids.

The carbohydrates remain undigested in the stomach. In the small intestine, the pancreatic enzymes (e.g., $\propto$-amylase) break carbohydrate and fat contents of processed food. The bile is produced in the liver and stored in the gallbladder. It is secreted into the small intestine for the digestion of fats and some vitamins. There are bacteria in the small intestine, which also help to digest the carbohydrates. The muscles contract and relax continuously to mix food with the digestive juices. Its wall absorbs water and nutrients such as amino acids, glucose, glycerol, vitamins, and salts to pass into the bloodstream to deliver them into the liver. The liver stores and processes nutrients that are then carted to the tissue cells all over the body, by the blood, for energy production, growth, and cell repair.

As peristalsis continues, the residue is pushed into the large intestine where some bacteria help it to break down into vitamin

K. The remining unprocessed wad consists of indigestible solids, fluid, and old cells from the lining of the GI tract. The wall of the large intestine absorbs water from the fluid waste to convert it into the solid stool and shove into the rectum. The rectum stores stool temporarily until a bowel movement pushes it out of the body through the anus.

The digestive system is also functionally connected with the other organ systems. Cells in the stomach and the small intestine produce and release hormones that help the digestion. Hormones also send signals to let the brain know when we are hungry or full. The central nervous system also controls some of the digestive functions. For example, when we perceive or get the aroma of food, the brain sends a chemical to trigger the salivary glands for secreting the saliva and make the mouth ready for eating. Nerves within the walls of the GI tract also release substances that control the movement of the food and the production of digestive juices. Also, those nerves send chemicals for helping the gut muscles in contraction and relaxation to propel the cud through the intestines.

Urinary system

As the digestive system produces solid waste, the urinary system produces liquid waste, called urine, from the remaining of the food waste and eliminates it from the body.

Figure 3.15 shows the urinary system, also known as the renal system, which contains kidneys, ureters, bladder, and urethra. The functional units of kidneys are called *nephrons* that produce urine by filtering food wastes and absorbing water from

the blood. Urine flows out of the kidneys through ureters and fills the bladder. When the bladder is full, urine gets eliminated through the urethra.

The urinary system also works with other organ systems such as lungs, skin, and intestines to maintain the balance of chemicals, water contents, blood pressure, and pH in the body. Blood reaches the kidneys through the renal arteries and goes away through the renal veins.

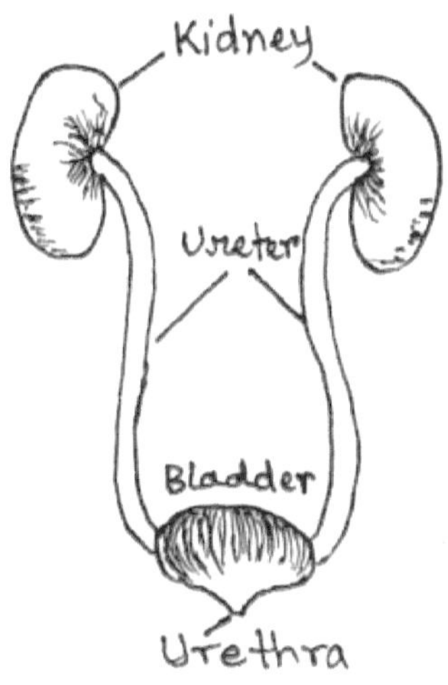

Figure 3.15. Human urinary system.

Our body has a reproductive organ system, which has not been taken in this book. All the organ systems interact together to maintain the *homeostasis condition* in the body. Therefore, when we take medicine, say for cold, it is not only the respiratory system but the other connected systems also do get affected. That is considered as the side effect of the medicine. The physical build-up of a person depends on many factors, e.g., inherited genes, environment, and culture, which affect the organ systems. Thus, those factors must be taken into consideration while medicating for a disease. We should always consult a specialist doctor for medication; because doctors always follow *The laws of medicine* (Mukherjee, 2015).

Medicine

The term medicine implies the subject of understanding disease and establishing methods of curing it through diagnosis, prognosis, and treatment. Pharmacy, closely related to medicine, is the subject of preparation, formulation, and dispensing of medicinal drugs. Pharmacology explains the sources, uses, and mechanisms of action of drugs. Disease, usually associated with some symptoms, is considered as a negative deviation from the homeostatic condition of a living body. This section gives a limited understanding of how medicines, i.e., drugs, do work inside the body. The drug is not always medicine. *All medicines are drugs, but all drugs are not medicines.* Medicine treats diseases; they have a positive effect on patients; but a drug can have either a positive or a negative consequence. Drugs with a positive outcome are medicines. However, drugs like heroin do not treat disease.

In this chapter, drug refers to medicine only. However, the medical benefits of the drug will assume if taken in the right dose, or else, it may harm like a poison. Swiss physician, alchemist, lay theologian, and philosopher of the German Renaissance,

Paracelsus said,

"Poison is in everything, and nothing is without poison. The dosage makes it either a poison or a remedy."

A drug is a chemical compound, which may be prepared either from herbal products or from synthetic chemicals in pharmaceutical laboratories. Drugs interact with the germs that invade our body, destroy them. Drugs also destroy harmful body cells (e.g., cancer cells) as well as supplement nutrients (e.g., hormones or vitamins) deficiency in the body.

How does a drug work in the body? It initiates a chain of actions. Suppose there is a pain somewhere in your body. So, you swallow a painkiller pill. The pill reaches the stomach through the food pipe, goes into the small intestine, and finally into the liver, where it decomposes into its constituent molecules. From the liver, those molecules get absorbed through the capillary blood vessels into the bloodstream, circulate throughout the body, and bind with membrane receptors at the diseased site due to "lock-and-key" interaction.

The cell membrane has two types of integral proteins: receptors and effectors. The receptors, which act like the sense organs of the cells (Lipton, 2018), project out like antennas for receiving the chemical signals. Each antenna has a specific chemical structure that detects only a single type of chemical communication. Thus, drugs are designed with specific chemical agents so that they can target the receptors of the targeted cells to implement the remedial benefits. When we take ibuprofen ($C_{13}H_{18}O_2$) or paracetamol ($C_8H_9NO_2$) for pain, the drug molecules circulate throughout the body with the bloodstream. So, wherever they find respective receptors generated by the painful inflammation, in the cell membrane, bind there. The chemical agents used for targeting drugs are called ligands. The ligand-receptor binding is the shape matching "key-in-lock" type, i.e., the ligand is like a key that searches for the lock (receptor) where it fits.

When a tissue is injured, its constituent cells send a chemical message, like an FIR (first incidental report), through the neural network to the brain and triggers the consciousness. Needless to say that the messaging process is intricate. The injured or infected cell, first, releases a chemical called *prostaglandin*, which is sensed by the nerve synapses. Then they send the message (by triggering the action potentials across a series of neurons) to the brain. The brain immediately sets off an awareness (i.e., prompts consciousness) about the injury to feel pain in the affected area, so that a necessary action, such as gulping a painkiller pill, is taken.

The drug molecules travel through the bloodstream to bind at the injured cell membranes and inhibit the prostaglandin secretion. Thus, the synapses duly stop sensing the prostaglandin. Subsequently, the neurons stop stimulating the action potentials across the synapses, so that no message goes to the brain, and we feel relief. Pain is a warning signal for an injury or infection; ignoring it may cause acute disease. The process of the inhibition of prostaglandin production is as follows.

The phospholipid cell membrane contains polyunsaturated fatty acids. Those are called arachidonic acid ($C_{20}H_{32}O_2$). Infected or injured cell membranes release arachidonic acid (AA) into the

cell. There it binds with cyclooxygenase (COX) enzymes and catalyzes their oxygenation. The oxygenated COX produces the prostaglandin to initiate the inflammation. The molecules of a painkiller drug bind at the active sites of COX and block it from interacting with the AA molecules and, thus, stop the production of prostaglandin.

Almost all human cells have the mechanism of generating prostaglandin, which may have different chemical forms. It is the most potent chemical that creates the inflammatory response (e.g., pain, swelling, fever, redness) against an injury or disease as the first step of the healing process. If a blood vessel is injured, its cells produce a type of prostaglandin called *thromboxane*. It creates an inflammation and, simultaneously, stimulates the formation of a blood clot with the muscle contraction in the blood vessel to prevent blood loss. Afterward, another type of emancipated prostaglandin called *prostacyclin* stops clotting. It also prompts muscles in the blood vessel to relax and increase the blood flow. Thus, the response to the injury and the inflammation gets reduced. Prostaglandin also causes an inflammatory response against pathogens (viruses and bacteria) attack, which triggers the production of the T-cell such as neutrophil. Neutrophil, filled with tiny sacs of the enzyme, eats and digests pathogens.

Histamine ($C_5H_9N_3$) is another signal molecule. It turns on the inflammatory response such as itching, sneezing, runny nose, itchy eyes, congestion, and sinus pressure against some allergies such as hay fever and the common cold. For the relief, we ingest antihistamine drugs such as Cetirizine, Allegra, or Benadryl. It gives us comfort but may cause one or more other symptoms such as drowsiness or sleepiness and blurry vision. The reason is as follows. During circulating through the bloodstream, the drug can bind with all cell membranes wherever found with histamine receptors. As the drug molecules are very tiny, they can even cross the blood-brain-barrier (BBB) and enter the brain; subsequently, affect the neural cells there. As a cause, the neural signals alter. That affects the nervous system causing drowsiness and other symptoms. Consider another example. When a chemotherapy drug ingress into the bloodstream of a cancer patient, its molecules search for fast-growing or rapidly dividing cells, as malignant cells have those characters. But hairs and nails are also fast-growing, or rapidly dividing cells, which also are damaged by the chemotherapy drug. Thus, patients lose their hair.

As organ systems in the body are functionally interconnected, there is a continuous flow of information from one organ system to the other, creating a *holistic* network (Lipton, 2018). But the holistic physiological condition differs from person to person. For that reason, the same drug does not benefit everyone the same. Hence, the physiological state of a person needs to examine first, using the laws of medicine, before prescribing a drug.

The basis of the laws of medicine includes *priors, outliers,* and *biases* (Mukherjee, 2015). Prior means the prior knowledge about the patient, i.e., his or her family history of health and diseases, past and present neighbourhood, culture, and many other factors that influence the physical and mental condition of a patient. According to the Cambridge dictionary, the outlier differentiates between a person, thing, or fact from another person, thing, or the actuality, and restricts a general conclusion between them.

So, the prior knowledge of the family history of a patient may not tally in all cases. The family history of health predominantly flows through genes. There are around 20,000–25,000 coded varieties in the human body. Therefore, a mutation in the coded sequences of merely a few genes makes an individual lie outside the family history. Bias is the biasness of individuals, including both patient and the doctor, to certain beliefs, thoughts, principles, tests, names, dates, times, and many other factors that affect the treatment process. Unfortunately, we still ignore those laws of medicine. Voltaire wrote,

"Doctors are men who prescribe medicines of which they know little, to cure diseases of which they know less, in human beings of whom they know nothing."

Fortunately, there is a wake-up! Our thinking process is taking a new path. Future medicine will come after tailoring to the gene level. Thus, a patient would get a DNA test first (Davis, 2006). In another few years, scientists expect that the study of genes (that includes how genes influence actions, appearance, and health) will dominate medical therapy. The prescription of medicines would follow the chemical needs in the body of a patient. It is called the *pharmacogenetic* approach. Besides, researchers are working on smarter methods for the targeted delivery of drugs to reduce the side effects drastically. Targeted heating (known as *hyperthermia*) is also an intelligent therapy for killing affected tissues like cancer tissues. Research in this direction requires knowledge in nanoscience. According to the American food scientist and microbiologist, Samuel Cate Prescott,

"Until that happens, we'll have to live with dumb drugs."

Present methods of medicine work on the concepts of Newtonian mechanics, according to which an organ is a massive object, built of atoms, and the interactions in the body are local at specific areas. So, the effect of medicine has to be confined, which is fallacious. According to quantum mechanics, *an atom has no physical structure; it is a vortex (e.g., see Figure 1.7) of energy that is constantly spinning and vibrating*. A physical body is built-up of non-localised energy (Lipton, 2018), which is known as *vital energy*. As we know, the relationships between mass (m), energy (E) as well as wavelength (λ) is $mc^2 = E = hv = hc/\lambda$. Thus, the body is a composition of energy waves where the organs are like vortices. Thus, the signal transmission between different organs and the brain takes place through the energy-wave propagation that is instantaneous. An individual human body radiates a unique energy signature that is useful for the diagnosis of a disease by analysing the energy spectrum. Such an energy plot is obtainable from a modern energy scanning device such as a CAT scan, MRI, and PET scan. But, are those measurements very accurate? Heisenberg's uncertainty principle implies that an energy spectrum is bound to a certain amount of uncertainty. The uncertainty relationship is $\Delta E \Delta t \geq h/2\pi$, where ΔE and Δt are uncertainties in the measured energy and time, respectively. When energy wave (e.g., light, sound, or magnetic field) is the probe, the measured spectrum owes to have a certain amount of uncertainty or inaccuracy. But that will be negligibly small compared to several advantages of those measurements. Besides, the energy radiation has therapeutic applications. For example, it makes constructive interference between the probing waves and the waves from the kidney by phase-contrast matching. It generates localised heat for exploding and dissolving stone in the kidney (Lipton, 2018).

Radiation therapy can also kill cancer cells. The mental disorder is also sometimes treated with shock wave therapy by sending electrical pulses in the brain.

Professor Amit Goswami retired from the Department of Physics, University of Oregon, USA, wrote about the Integral Medicine (IM) in his book, entitled, *Quantum Doctor: A Physicist's Guide to Health and Healing*. What is Integral medicine? It is a combination of conventional allopathy medicine and alternative medicines. Alternatives are homeopathy, Indian Ayurveda, and Chinese *chi* therapy, such as acupuncture. The Integral medication uses the quantum (i.e., non-local) communication of electrochemical signals. Alternative medicines prefer mind over body, and also takes care of subtle energy (called *prana* or *chi*) and non-physical spirit (called consciousness). According to this school of thought, the human body has entanglement with the consciousness waves of the universe. Thus, it is getting influenced by nature. As mentioned in the Introduction, Scientists now believe that universal consciousness controls all structures and physical laws. Scientists relate the consciousness wave of the universe with the gravitational wave that got generated after the Big Bang. I mentioned the bottom-up steps of the formation of the human body; those steps get redefined now as:

Atoms → biomolecules → organelles and cells → tissues → brain (organ) → consciousness

Consciousness does control our physical body; it catalyzes the interaction between the mind and the body and accelerates the body's healing processes. As the conventional allopathy medicines are designed for local action only, they are incapable of taking care of the non-local holistic character of the body. In that sense, alternative medicine is way ahead.

Ancient Indian yogis did realise a connection between the mind and the physical body, and conscious practices of that connection keep us healthy. Thus, they suggested exercises to integrate the body's vital energy, which is known as *Yoga* in Sanskrit, which means *to join* or *to unite*. On request from Indian premier, Narendra Modi, UN General Assembly has declared 21st June as *Yoga Day* to motivate people across the world to practice yoga. Indian author-cum-spiritual master, Amit Ray said,

"Yoga is not a religion. It is a science, a science of well-being, a science of youthfulness, a science of integrating body, mind and soul."

CHAPTER FOUR

NANOSCIENCE

PURPOSE OF THIS CHAPTER: To know the role of the fundamental forces in the nanoscale regime of matter.

The elementary particles like electrons, quarks, and gluons got formed from the energy released in the Big Bang. At some lower temperatures of the universe, the elementary particles confound due to the strong nuclear force into bigger particles like protons and neutrons. The same enforcement further caused the formation of nuclei of light elements like hydrogen (H) and helium (He) by confining protons and neutrons. At much lower temperatures of the universe, the electrostatic force also became prominent. This new force brought the negatively charged free electrons in the proximity of the positively charged nuclei. Those small united entities shaped the new fundamental particles, called *atoms*, like H and He.

A much weaker gravitational force came into action when the universe became more chilly. This force drove atoms and dust to cluster into various physical forms of size, starting from microscopic scales to even cosmological bodies like planets, stars, and galaxies.

The nuclear forces are very short-ranged; it is the most effective only within the size of a nucleus. On the other hand, electromagnetic and gravitational forces are long-ranged. Nanoscale sciences have mostly benefitted from electromagnetic energy. In this regime, different branches of science such as physics, chemistry, biology, medicine, and engineering have culminated. One nanometer (nm) is equal to 10^{-9} meters, and nanoscience is the exploration of physical phenomena of matter within such a tiny length scale, which is not an easy task. It requires many sophisticated techniques as well as advanced knowledge.

Like any other branch of science, nanoscience also has a beginning, which had bloomed in a great mind. On December 29, 1959, Professor Richard Feynman delivered a lecture at the annual American Physical Society meeting at Caltech. The title was *There's plenty of room at the bottom: An invitation to enter a new field of Physics*, where he suggested exploring science within the nanoscale.

Let me take some examples to give a feeling about the nanometric length scale. A paper sheet is about 100,000 nm thick; the average diameter of human hair is about 80,000 nm. The size of some biomolecules such as DNA is about 2.5 nm, and the folded protein globule is about 10 nm. The size of a red blood cell (RBC) is around 7000 nm. Figure 4.1 shows the size of some biomaterials. The selected length scale is from 100 picometers (1 pm = 10^{-12} m) to 1.0 millimetre (mm).

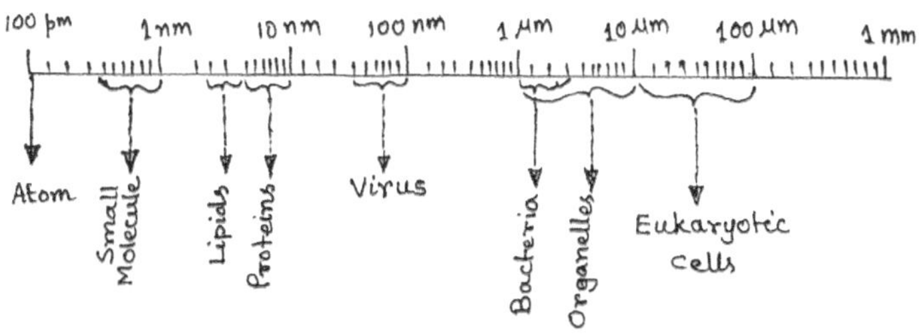

Figure 4.1. Some biomaterials on different length scales.

NANOMEDICINE: The application of nanoscience in medicine is called *nanomedicine*. The main focus in this area is to incorporate nanomaterials in the diagnosis and therapy of diseases at the targeted sites. The best size of nanomaterials for such applications varies between 10 to 100 nm, as particles in that size range can reach anywhere in the body. Nanomaterials bestow many applications in medicine. For example, imaging and biosensing, targeted drug or gene delivery, and hyperthermia. Also, nanomaterials are in use in many domestic applications such as laundry and cosmetics, in industrial applications such as underground oil recovery and the latest electronic appliances.

FORMATION OF NANOMATERIALS: Most of the natural materials form through *bottom-up* processes. For example, from elementary particles like quarks and electrons to nucleons (neutron and proton) to atoms, then to microscopic objects, and finally to macroscopic objects as solids, liquids, and gases. Each of those steps is meticulously designed by the fundamental forces. Consider a piece of ice of size 1.0 cm^3. It is an assembled structure of a large number ($\sim 10^{12}$) of microscopic ice crystals, and each ice microcrystal forms with a large number ($\sim10^5$) of water molecules (H_2O). A water molecule composes out of two H atoms and one O atom. An atom is a bound unit of electrons and nucleus, where the core is a composite of neutrons and protons. A proton or a neutron composes of quarks and gluons. Those bottom-up steps start with the elementary particles like quarks, gluons, and electrons and then proceed step-by-step to constitute an ice crystal. A reverse process that starts from the top, i.e., from the ice crystal and then returns step-by-step down to elementary particles, is called *top-down* steps. Nanomaterials can form in both ways. However, such tiny material can contain only a few numbers of atoms. Thus, nanomaterials have very exotic physical properties. According to Professor Richard Feynman, the fundamental question is,

"What would happen if we could arrange the atoms one by one the way we want them?"

In the bottom-up processes, matter gets formed through the chemical interaction between several atoms, either upon sharing (covalent bonding) or exchange of electrons (electrovalent bonding). In a chemical reaction, raw salts are the precursors. In the first step, bonds of the precursor molecules break, and their constituent atoms get isolated. Those atoms then get rearranged one by one, 'the way we want them' by forming new bonds, which give new chemical products. In this way, nanomaterials get formed, in most cases, which is called chemical route synthesis. In the top-down process, a solid with the required chemical composition gets machined down, 'the way we want' using the mechanical force, to nanometer size. A ball-mill can produce nanomaterials from bulk solid.

Due to their small size, the surface-to-volume ratio of nanoparticles is very high. This factor makes them highly reactive and toxic for health; it is technically called *nanotoxicity*. The tiny size also renders exotic properties to nanomaterials. An increased electrical conductivity in nanowire is usual than its bulk metal. Magnetization also increases in magnetic nanoparticles. Again, we can see an enhanced optical absorbance in nanoparticles compared to the same metal solids.

Nanotoxicity

All of us have seen the curve surface of the water in a glass container, or the spherical water droplets. Curvature increases if the size of a material decreases, i.e., the curvature varies inversely with the size of an object. Nanotoxicity arises mostly because of the curvature of the nanoparticles.

Why does curvature increase with decreasing size? Let us consider the example of the water surface. Water molecules (H_2O) are composed of H and O atoms through covalent bonds. Inside the liquid, none of those bonds is free or dangling. However, on the surface, there are many dangling bonds. Those free bonds retain high energy and look for a partner to distribute the same. That situation with high surface energy incorporates stress at the water surface, called the *surface tension*. To reduce the surface tension, dangling bonds of water molecules at the facet move towards the glass wall to share their energy. Thus, the water surface gets pulled and becomes curved near the glass wall. The surface tension is an inherent property of materials in both solid and liquid forms. It develops further if the surface area increases to a given volume.

Thus, the *surface-to-volume ratio* accounts for the surface tension in a material. As a solid surface can't change its shape, it will have more surface energy, so more surface tension, compared to a liquid.

Due to tiny size, nanomaterials have a very high surface-to-volume ratio, as well as surface energy. A simple calculation can explain this. Let us take a cube with sides 'a' as shown in Figure 4.2 (a), and cut it into eight identical cubes with sides '$a/2$', as shown in Figure 4.2 (b). The volume ($V1$) of the original cube is equal to a^3, and surface area ($A1$) $6a^2$. Each small cube has a volume $a^3/8$, and surface area $3a^2/2$. As there are eight small cubes, the total volume of newly formed cubes is $V2 = 8{\times}a^3/8 = a^3 = V1$. Thus, by reducing size, the total volume does not change. However, the sum of the surface areas, $A2 = 8{\times}3a^2/2 = 12a^2 = 2{\times}A1$, has become double; or, in other words, the surface-to-volume ratio (S/V) has increased by a factor of 2. The ratio S/V increases if the particle size gets further reduced. Due to their tremendous surface energy, nanoparticles are very reactive, and consequently, highly toxic for health applications. To reduce the toxicity, usually, a coating of inorganic or organic molecules is made on the nanomaterials. The surface coating also prohibits the clumping of nanoparticles in the liquid medium and the dispersion stable.

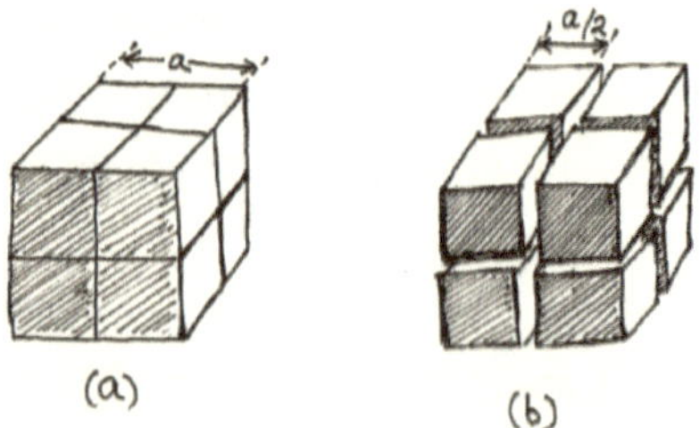

Figure 4.2. Cubes of (a) side 'a', and (b) side $a/2$.

Electrical property

When atoms locate close, their orbital energy levels interact and slightly alter that create energy bands in rigid bodies, as shown in Figure 1.22. In crystalline solids, there is a *valence band*, a *conduction band*, and a gap between the two, which is called the *energy gap* or *bandgap*, where no electron can exist. The valence band electrons can gain energy and jump to the conduction band that can generate an electrical conduction current in the solid. As shown in Figure 1.23, metals have either zero or very small bandgap causing high electrical conductivity. In semimetals, the gap is more than metals but less than 3 eV, and in insulators, it is more than 3 eV. The energy required by the electrons to transit from the valence band to the conduction band can have in the form of voltage or heat or light from an external source. The width of energy bands and bandgap depends on the number of neighbouring atoms in a crystal, and it is an intrinsic property of a material. If the crystal size reduces, the number of nearest-neighbours also reduces that narrows down the bandwidth and, in turn, widens the bandgap in semiconductors or insulators, as shown in Figure 4.3. It is called the *quantum size effect*. It may convert a semiconductor into an insulator. Consequently, the electrical conductivity of semiconductors reduces. As conductors do not have a bandgap, the reduced width of the valence band causes more electrons to transit to the conduction band and, thus, increases the electrical conductivity.

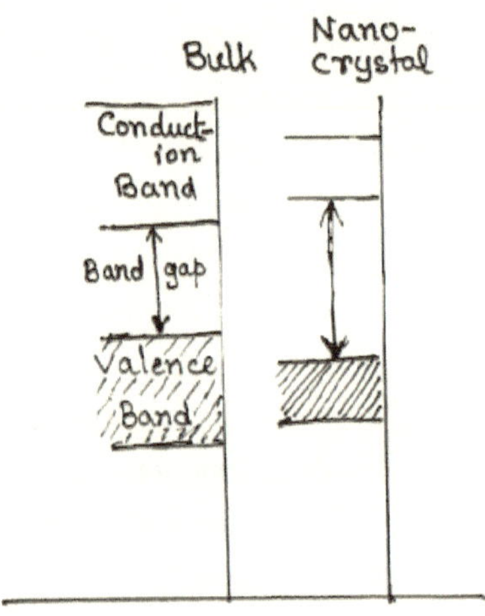

Figure 4.3. Band structures of bulk and nanocrystalline materials.

Magnetic property

The crystalline solids could be either magnetic or nonmagnetic. Magnetic solids such as iron get attracted to the magnet, whereas nonmagnetic solids like aluminium don't. A magnet has a north (N) pole and a south (S) pole. Magnetic solid consists of a large number of tiny magnets with their N and S poles randomly oriented, resulting in no net magnetism. Since each mini magnet has two poles, it is called a *dipole*. When the same solid comes within a magnetic field (say, due to a permanent magnet), its tiny magnetic pieces align in such a fashion that the solid temporarily acquire magnetism. It is called *induced magnetism*. The exposed solid provisionally becomes capable of

attracting other magnetic materials.

Why the tiny magnets in the solid align? First, we should know why those mini magnetic entities are so randomly oriented. At standard temperature, those building minuscule blocks gain high thermal energy as $E = k_B T$, where k_B is the Boltzmann constant, and T is the temperature in Kelvin. There is a characteristic temperature for all solids, called the *Curie temperature* (T_C). Above T_C, the thermal energy of their constituent dipoles is higher than their magnetic exchange interaction energy; thus, the tiny dipoles remain random.

Below T_C, the magnetic energy becomes higher, and the *dipole-dipole interaction* is set-up. The dipolar interlinkage energy is $U_{dip} = -2p^2 cos\theta / 4\pi\varepsilon_0 r^2$, where p is the dipole moment of each tiny magnet, is the permittivity of vacuum, r is the distance between the centres of two neighbouring dipoles, and θ is the angle between the axes of two tiny magnets along their lengths. This dipole-dipole interaction causes pair-wise magnetic attraction below T_C. Above T_C, the application of an external magnetic field (H) causes each dipole to interact with it. That dipole-field interaction is attractive with energy, $U_{df} = pHcos\theta$. In that situation, the net conforming energy in the magnetic solid becomes, $U_{net} = (U_{dip} + U_{df})$. If H is large enough so that $U_{net} \gg E$, magnetic dipoles in the solid attain the orientation so that $cos\theta$ becomes unity. That makes the system achieve a maximum attractive situation, i.e., θ has to become zero. The maximisation of the U_{net} causes the alignment of the tiny dipoles along the direction of the external field.

On withdrawal of the external field, the U_{df} vanishes, and the thermal energy dominates to again randomising the dipoles' directions. But it is a little sluggish, and the solid retains a fraction of the induced magnetisation for a while. It is called *residual* or *remnant magnetism* (M_{rem}). The capacity of a solid holding its induced magnetism is known as *retentivity*. It is an intrinsic property of the material. What will happen if, instead of withdrawing, the direction of the external field reverses? All tiny magnets in the solid also slowly reverse their alignment, to re-establish the negative maxima of U_{net}. In that case, the net magnetization gets reversed. Such back and forth change of the external field creates a loop in the magnetization (M) versus magnetic field (H) plot, as shown in Figure 4.4 (a). It is called a *hysteresis loop*. It reveals the magnetic-character of a solid. For example, iron is characterised as soft or hard magnetic if the loop width is narrow or broad, respectively.

In the case of magnetic nanoparticles, the hysteresis loop does not exist, i.e., no remnant magnetism at zero external fields, as shown in Figure 4.4 (b). Because of their tiny size, nanoparticles have substantial thermal energy for which $E \gg U_{dip}$. The U_{df} becomes zero as soon as H disappears, the dipole-dipole coupling breaks, and the magnetic nanoparticles lose their magnetism. This fact makes nanoparticles suitable for medical applications, as the remnant magnetism may cause clustering of nanoparticles and retain in the body for a longer time leading to toxic effects.

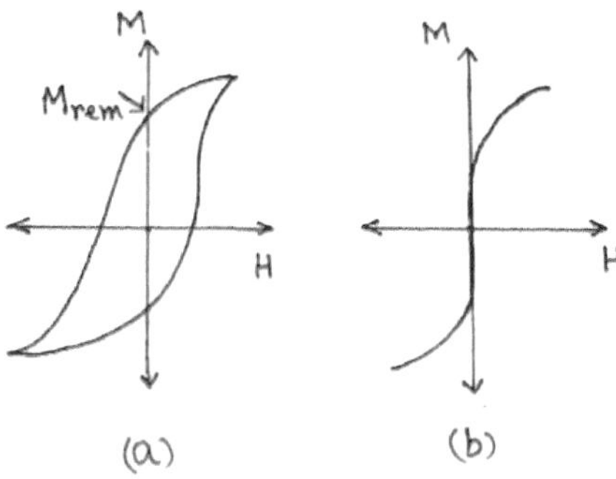

Figure 4.4. M-H plots for magnetic (a) bulk solids with M_{rem} as the remnant magnetism, and (b) nanoparticles, with and without 'hysteresis loop', respectively.

Optical property

Metals have a large number of loosely bound conduction electrons at the surface, execute random oscillations. Upon absorption of energy from incident white light (contains seven colours, VIBGYOR), the surface electrons perform coherent oscillations with the frequency of absorbed photons. Consequently, a resonance, known as *surface plasmon resonance* (SPR), will generate. After putting out the incident light, the electrons emit radiation at a specific frequency to come back to their ground states. Such emission of radiation produces an SPR band, which reveals the optical characteristic of a metal. For example, the silver film generates the SPR band around 400 nm.

As the metal nanoparticles have a higher surface-to-volume ratio compared to a bulk metal, they emanate more intense SPR bands. The SPR band is sensitive to the surrounding medium in contact with the metal surface. The band modifies upon a small change in the medium. In this regard, nanoparticles are more sensitive than thin films or bulk. This property of nanoparticles is especially useful for designing of the colour-based optical biosensors. The frequency or wavelength of the SPR band varies with the nanoparticle size. Thus, nanoparticles have some exotic properties which offer applications in technology and medicine.

Medical applications of nanomaterials

Why do we need nanomaterials in medicine? The current methods of the delivery of drugs are crude! Usually, we take oral pills or intravenous injections. Once the drug molecules enter into the bloodstream or body fluids, reach everywhere in the body and cause some unwanted side-effects. Moreover, due to the dilution in the body fluids, excess quantities are required to intake. Therefore, to minimise the chemical intake delivery of the drug at the affected sites is a meaningful approach. How to achieve such a drug delivery method? It could be the cell-membrane receptor-based targeting, and it is the most practical way as of now. But that does not reduce the risk of side effects. For example, chemotherapy for a carcinogenic tumour causes hair-fall or destruction of some healthy cells in the body. In fact, in ample cases, after-surgery chemotherapy becomes the cause of mortality of cancer patients. Unfortunately, chemotherapy is one of the viable treatments for cancer today. So, we need more precise targeting of drugs, and nanomaterials bring some hopes in that direction.

Nanomaterials have several medical applications such as

detection and diagnosis of diseases, targeted therapies, imaging, and drug or gene delivery. However, for any specific application, nanomaterials need to be coated with particular *functional molecules*. For instance, targeting a tumour, nanoparticles need to be conjugated with the antibody "key" that can bind only to the cell-membrane receptor "lock" at the targeted site. The human cells have an intelligent way of expressing specific receptors (keys) in the cell-membrane at its dysfunctional state. Thus, those nanoparticulates are called *functional nanoparticles*.

Nanomaterials are synthesized through some bottom-up steps, usually, in any chemical route. For *in vivo* applications, nanoparticles must have a size range within 50–100 nm.

Figure 4.5 shows the cartoon of a functional nanoparticle. The core contains the spherical nanomaterial and the shell, the utility coating of organic or inorganic molecules. Nanomaterial could be made of either natural or synthetic polymers (e.g., albumin, chitosan, or polystyrene), or metals (e.g., iron, silver, or gold), or metal oxides (e.g., ZnO, Fe_2O_3 or Fe_3O_4). The coating molecules could be polymers such as poly(ethylene) glycol (PEG) and dextran or surfactants to passivate the surface. Passivation helps to prohibit the clustering and toxicity of nanoparticles. Such coating is possible through either hydrophobic, or van der Waal, or electrostatic interaction between the nanoparticle surface and the coating molecules.

For drug delivery or radiotherapy or endoscopy at specific tissues, an overlay of additional functional species such as drug or radioisotopes or fluorescence dye is necessary, in addition to the specific antibody-coating. Unlike the protective coating, application molecules do not bind directly to the nanosurface. For this purpose, different binding chemistry becomes required. Usually, the linker molecules provide the necessary chemistry through electrostatic interaction. One end of the linker binds with the nanoparticle surface and the other end with the functional species. There are versatile epoxy amine linkers, which can link different types of working molecules with nanomaterial surfaces (Nickels et al., 2010). One end of those linkers contains an epoxide terminal, and the other end an amine terminal.

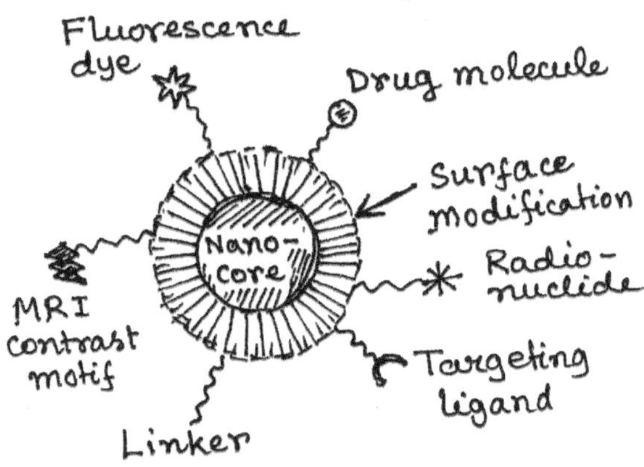

Figure 4.5. Cartoon of a functional nanoparticle.

Diagnosis

As for applications, functional nanomaterials have shown promises both in diagnosis as well as therapy of diseases. In this section, I shall discuss applications in two diagnostic techniques, namely, (i) *biosensor* and (ii) *magnetic resonance imaging* (MRI).

BIOSENSOR: A biosensor is an analytical device that can detect disease by testing a sample such as blood, urine, or saliva collected from a patient. Those samples would contain either specific cells, molecules, genes, proteins, enzymes, or hormones, called *biomarkers*, which get expressed in the cells with disease-specific signatures. For example, proteomic biomarkers are specific proteins that get overexpressed in the cells or cell membranes on acquiring a disease. Biosensors have a biorecognition element that can detect the disease-specific biomarkers in the patient's sample and help in diagnosis. The general structure and the working principle of a biosensor have briefly discussed in the following section; details are available elsewhere (Inamuddin et al., 2019).

A sketch of the basic structure of a biosensor has shown in Figure 4.6. It contains a *bio-recognition component*, coated with specific receptors such as enzymes or antibodies, which detect biomarkers in the sample. The binding of biomarkers with the biorecognition elements follows the shape-matching "lock-and-key" mechanism, as shown. It gets realised through the electrostatic interaction between the atoms of biomolecules and analytes.

The biorecognition component communicates the binding interaction to a *transducer* to converts it to a readable output, such as the change in the voltage or current or the intensity of the optical spectrum for further processing and analysis.

The transducer output is first amplified and then digitised using the microelectronics circuit before sending it to the output device. The output signal helps to identify and assess the disease and prognosis.

The sensitivity of a biosensor depends mainly on the biorecognition component and the transducer (Leca-Bouvier and Blum, 2005). There are different types of biosensors (Ghosh, 2019).

A conventional biosensor can diagnose a disease unambiguously if the signal strength from the biorecognition element is strong. It is possible if there are a good number of biomarkers in the patient's sample. But that would be an advanced stage of the disease like cancer. Starting therapy at an advanced stage of the disease usually does not benefit. This drawback can get the better of on increasing the surface area of the bio-recognition component without actually increasing its size. A layer of nanomaterials on a suitable substrate can facilitate a large surface area, due to the large surface-to-volume ratio of nano-particulates, for attaching more number of receptor molecules than on a flat surface.

Alternatively, functional nanoparticles may be directly deposited on the transducer as the biorecognition film with an enhanced area to facilitate an increased number of biomarkers binding. Consequently, a remarkable chemical signal would generate from the biorecognition component, and the sensitivity of the biosensor would increase by several folds.

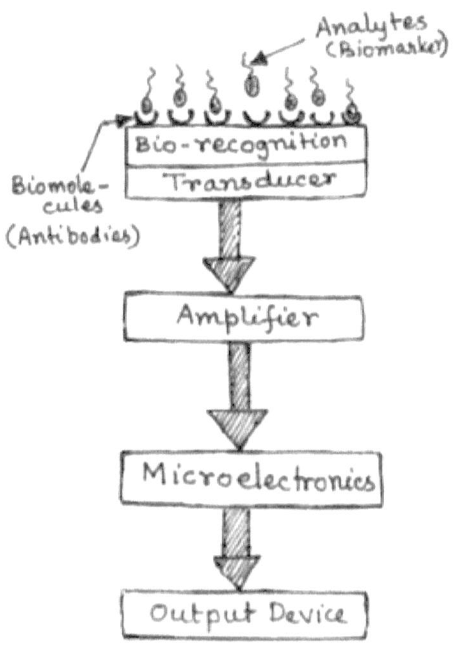

Figure 4.6. Schematic diagram of the basic structure of a biosensor.

Nanoparticles can also inherently generate a readable signal such as SPR spectrum or magnetic field, upon application of an appropriate probe, and can work as the transducer. For example, metal nanoparticles (mNPs) such as silver or gold nanoparticles can produce an intense SPR signal upon the incidence of the visible light. If biomarkers bind on mNPs, its original SPR signal gets modified, which confirms the presence of the biomarkers in the sample. The SPR signal from mNPs can get further enhanced if alkaline-metal ions are present on the nanoparticle surface, as my group has shown on silver nanoparticles (Ghosh et al., 2018).

The presence of positive ions changes the dielectric character of the surrounding medium that creates more surface electrons of the mNPs and, consequently, enhances the SPR signal. Such mNPs can even supplement the use of a signal amplifier. These ways, the basic structure of a biosensor becomes simpler and smaller.

MAGNETIC RESONANCE IMAGING: It is an imaging technique that gives the image of an organ such as the brain to show the presence of a tumour or other disease. In this imaging, RF radiation is allowed across the affected area of the patient's body. The emitted ray excites the proton spins of H atoms in the water (H_2O) content at the specific tissues and probes the changes in their spin alignment. Before the RF irradiation, the patient gets exposed to a strong magnetic field for aligning those proton spins through the magnetic dipole-dipole interaction. Application of RF field excites H atoms to misalign proton spins so that after turning it off, spins relax to realign along the applied magnetic field. While slackening from the excited to the ground state, H atoms give out energy, which generates the MR image of the targeted organ. Physicians analyse the MR image and assess disease progress or prognosis. For an authentic assessment, the image must have a good contrast; a blurred slide often leads to wrong analysis.

The contrast of an MR image depends on the relaxation-rate of the proton spins; a faster relaxation enhances the contrast. As in some cases, the image appears unclear due to a weak resolution, the physician injects a contrast-enhancing agent intravenously in the patient's body a few hours before taking the scan. Usually, a gadolinium-based contrast agent is used, which accumulates in the targeted organ and increases the rate of relaxation of proton

spins. Gadolinium is a paramagnetic element of a high magnetic moment. It acquires magnetism in the presence of an external magnetic field that builds up the net magnetization in the tissues of the targeted organ. With a higher magnetic strength, the proton spins realign along the external magnetic field at a much faster rate as soon as the RF field disappears. Consequently, a contrast-enhanced MR image appears. Gadolinium preferentially shortens the longitudinal relaxation time (T1) of H atoms.

Despite having higher magnetic-moment ferromagnetic materials are unused as contrast agents, as they may retain magnetism even after switching off the external magnetic field. The residual magnetism may affect the structures and functions of the iron-bound proteins such as haemoglobin and ferritin in the blood.

Magnetic nanoparticles (MNPs) are superparamagnetic. Thus, they have a high magnetic moment but no magnetic retentivity. Moreover, each nanoparticle contains many iron atoms. So, it has a higher magnetic moment than a single gadolinium atom. Even that could be higher than a few linked gadolinium complexes. Also, due to a significant magnetic moment, a smaller amount of MNP based contrast agents can produce better contrast than a gadolinium-based agent.

Therapy

Let us consider two therapeutic applications of the functional nanoparticles, (i) *targeted drug delivery* and (ii) *hyperthermia*. In both cases, specific functional nanoparticles are injected into the patient's body to accumulate in the targeted cells for necessary therapy.

TARGETED DRUG DELIVERY: For the targeted drug delivery, drug molecules are attached to the surface-passivated nanoparticles through the linkers along with the tissue-specific targeting ligands, e.g., antibodies. Sometimes, the drug is encapsulated within a nanocage, such as vesicles, to deliver at the targeted site. Such encapsulation is possible using the hydrophobic interaction between the drug molecules and the polymer or lipid molecules of the cage.

At target-sites, nanocages face changes in the physiological conditions such as temperature or pH and open up. That causes the drugs to get off (Wilczewska et al., 2012). For the same reason, drug molecules attached to nanoparticles detach at the target site. There are several advantages of such targeted drug delivery over the conventional method. Because when the drug reaches directly at the desired locations, without interacting anywhere else in the body, the unwanted side effects will come down to a minimum. It also means a lesser amount of drug will be necessary for the purpose than in the conventional method. Though we need to travel a long way before getting the nanoparticle-based therapy available, the hope widens when a billionaire American businessman, Bernard Marcus, says:

"Nanotechnology in medicine is going to have a major impact on the survival of the human race."

HYPERTHERMIA: Application of nanomaterials, by and large, magnetic nanoparticles (MNPs), in hyperthermia, shows a great promise in drugless cancer therapy. In this execution also nanomaterials need to target the desired site. Furthermore, the toxicity of nanomaterials must be zero before a clinical trial.

Hyperthermia is a thermally driven method of killing targeted cells by heating them above 42 °C. When MNPs accumulate in those compartments, confirmed by endoscopy, RF pulse is allowed to pass through those areas. MNPs absorb energy from the radiation and alter their magnetic spins. While easing (called Neél relaxation), after switching off the RF field, nanoparticles release energy that heats the local site beyond 42 °C. This consequence the death of the cells, such as carcinomic tumours.

This method could substitute chemotherapy in the future. Chemotherapy has lots of side effects such as loss of hairs as well as body weight, or even damaging of healthy organs. Many times, those side effects push the patient towards death earlier than cancer.

Targeted ion-therapy using nanomaterials may also become a potential alternative medicine in the future. Investigation in this direction is in progress (Ghosh 2016). In this case, the primary role of ions would be to disintegrate the cell-membrane. That would ensure the perish of the tumour. Chemotherapy does the same but with lots of side effects.

Nanomaterials have created many new sciences and the possibility of several applications, starting from domestic utilities to advanced technologies and medicine. A few super minds like Professor Feynman could throw light on new aspects of science, which led to this new era of nanoscience. We have to keep in mind that nanoscience also depends on the fundamental forces. American actor William Powell said,

"Nanotechnology is manufacturing with atoms."

CHAPTER FIVE

ENERGY-MATTER DUALITY

PURPOSE OF THIS CHAPTER: To know how the fundamental forces accomplish the transformation of matter to energy and vice-versa.

"Concerning matter, we have been all wrong. What we have called matter is energy, whose vibration has been so lowered as to be perceptible to the senses. There is no matter."—Albert Einstein

According to *mass-energy equivalence*, $E = mc^2$, the matter is energy. Do they play a *dual role*? Or, is matter an illusion, truth is the energy? According to the Collins dictionary, *duality* is a situation in which two opposing ideas or feelings exist at the same time. Let's believe that the duality between energy and matter exists in nature, and there is a relation between them. On every occasion in our life *vis-á-vis* physical growth as well as physical or mental work, energy drives the show. Thus, vigour and matter are entangled. We do exercise and spend energy to lose our body fat, unwanted weight. It can be verified mathematically using mass-energy equivalence and work-force relationship,

$$\textbf{Energy} = mass \times velocity^2\,(mc^2)$$

$$= Mass \times (distance/sec)^2$$

$$= Mass \times \left(\frac{distance}{sec^2}\right) \times distance$$

$$= Mass \times acceleration \times distance$$

$$= Force \times distance = \textbf{Work}$$

A wave has vivacity $E = hc/\lambda$, where λ is the wavelength, and a particle is a matter. Thus, the title of this chapter means the *wave-particle duality*, which is the fundamental concept of quantum mechanics. This duality relationship, proposed by the French physicist Louis de Broglie, is given by $\lambda = h/p$, where p is the linear momentum of the particle, and h is the Planck constant. If the particle is at rest, p is zero, and is infinity, which means that the wave loses its significance to the particle.

The concept of duality between material and non-material first introduced by the Greek philosopher Socrates (470–399 B.C.). He perceived non-material as a "form of thought" that he defined as the "soul" and the nature of the universe as duality. In contrast, another contemporary Greek philosopher, Democritus (460–370 B.C.), believed in the supremacy of materials. He coined the word *atom* as the undividable entity of the universe. So, according to Democritus, the only thing that mattered was the *matter*; however, to Socrates, the physical entity is a "crude shadow" of a perfect form, i.e., soul (Lipton and Bhaerman, 2017).

The duality debate got surfaced again in the 1600s when scientists were investigating the nature of light. In that regard,

Christian Huygens and Isaac Newton, respectively, proposed modern *wave theory* and *corpuscular theory*. Later on, scientists had conducted experiments to verify those theories. Some of those experiments, *viz.*, reflection or refraction of light, supported the wave theory, whereas others, *viz.*, Compton scattering experiment, supported the corpuscular (i.e., particle) theory of light. Later, Thomas Young's double-slit experiment discarded the corpuscular concept and administered for the wave nature of light. But, the famous Michelson-Morley experiment raised doubt about wave theory. Professor Niels Bohr expressed the *duality paradox* as a fundamental or metaphysical fact of nature.

In 1905, Einstein proposed the *photon theory of light* while explaining the photoelectric effect, which is the ejection of electrons from a metal surface upon irradiation of UV light. According to that theory, light consists of photons, a packet of energy $E = h\nu$, ν is the vibrational frequency of the photon. The photoelectric effect has proved that light has both particle as well as wave nature, i.e., it possesses a dual character. For submicron particles such as an electron, this duality is verifiable through experiments. However, the wavelength of macroscopic objects is too tiny to detect. Thus, we do not consider a visible article as a wave. Only if a macroscopic object travel with a speed of light (3×10^{10} m/sec), it would appear as a wave. In practice, no macroscopic body can ever achieve the speed of light. The wavelength of a moving electron of mass, m_e (= 9.11×10^{-28} g), and speed, v (= 5.97×10^{6} m/s), is calculated using the de Broglie equation as,

$$\lambda = \frac{h}{p} = \frac{h}{mv}$$

$$= \frac{6\,63 \times 10^{-34} J - s}{(9.11 \times 10^{-28} g) \times (5.97 \times 10^6 \frac{m}{s})} \left(\frac{1000\, g - m^2}{J - s^2}\right) \left(\frac{10^9\, nm}{m}\right)$$

$$= 0.122\, nm = 1.22 \times 10^{-6}\, mm$$

For a macroscopic body of one-milligram mass moving at a speed of, say, 0.001 m/s generates a wave of wavelength, λ = 6.63×10^{-16} nm, or 6.63×10^{-22} mm, which is very small. Remember that the visible light has wavelengths in the range of 400–700 nm, i.e., $4 \times 10^{-4} - 7 \times 10^{-4}$ mm, which helps us in seeing the objects, but light itself is invisible to us.

Why does UV-light cause the ejection of electrons from the metal surface? Upon incidence of UV-light on a metal surface, the outermost orbital electrons of the atoms at the surface absorb energy and become free from their parent atoms, as shown in Figure 5.1. It is the same phenomenon as explained earlier for a hot iron rod. Since Planck constant h (= 6.626×10^{-34} kg-m²/s) is small, the frequency (v) of the incident radiation must be high to emit out electrons from the atoms overcoming the Coulomb attraction of their nuclei. Hence, UV-light is necessary to produce the photoelectric effect.

After the Big Bang, the universe consisted only of energy, no matter. Subsequently, part of that brisk got transformed into materials. Those were the elementary particles such as electrons, quarks, and gluons, mentioned in Chapter One. At some lower temperature, elementary particles accumulated into subatomic

particles such as protons and neutrons. On the way of cooling of the universe, neutrons and protons got compacted into nuclei. Then nuclei pulled electrons and formed atoms. Those hierarchical formations became possible due to the emergence of fundamental forces like the strong nuclear force and the electrostatic force. Finally, atoms fabricated matter using bonds. But there many question marks about those earthly bodies.

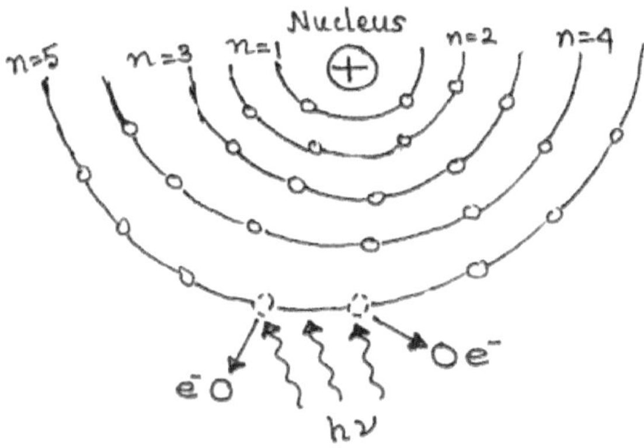

Figure 5.1. Cartoon showing the ejection of electrons (e^-) from an atomic orbit upon irradiation of light of energy, $E = h\nu$. The symbol Å represents the nucleus.

MANIFESTATION OF MATTER: Bombarding an accelerated electron with a high energy photon produces a low energy electron along with antimatter and energy. That experiment was carried out in the Large Hadrons Collider (LHC) in CERN at Geneva, Switzerland. The investigation aimed to find new, unstable, high-energy particles like the Higgs boson or the top quark that contributes to the *mass* of a matter.

What is the matter? According to classical mechanics, a material particle, or matter possesses a mass as well as size. According to *string theory*, point-like particles are one-dimensional *strings*, which appear as different particles when vibrated, twisted, or turned. For example, a vibrating thread of specific frequency and length behaves like a photon. The same is true for quark, gluon, or electron. Quantum mechanics considers particles as wave functions, and the existence of a matter gets realised through the collapse of the wave function that has the characteristics of that matter.

Do the elementary particles like an electron obey classical laws of motion? The laws of motion can calculate a particle's position and momentum very accurately. But the subatomic particles like an electron do not have a well-defined location or energy as recommended in the Heisenberg *uncertainty principle*. Those are quantum particles that hold a dual character, i.e., behave like waves as well as like particles. A quantum particle such as protons, neutrons, or electrons has a wave function giving a width in the probability distribution of its location. Quantum mechanics (QM) can calculate only the most probable placement of a quantum particle. Suppose a particle has a location x, in one dimension, at a given time t, then its wave function is written as,

$$\psi(x, t) = ae^{i(\omega t - kx)} \tag{5.1}$$

where $i\,(= \sqrt{-1})$ is a complex number. QM finds the probability of finding the particle between x and $x + dx$ at time t by integrating the particle's wave function with its conjugate wave function as,

$$\int_{x}^{x+dx} |\psi|^2 dx = \int_{x}^{x+dx} \psi(x,t)\psi^*(x,t)dx = 1 \qquad (5.2)$$

The finite value (normalized) of the integration indicates the collapse of the wave function. represents the complex conjugate of, i.e., . In quantum mechanics, the collapse of a wave function means the confining of a particle-wave at the most probable location using an external wave function. *Schrödinger's cat experiment* is a phenomenal thought experiment that explains the collapse of the wave function. Because of the *Copenhagen interpretation*[9], the German physicist and Nobel Laureate Erwin Schrödinger had designed it as follows:

A cat is placed inside a closed box along with a Geiger counter (GC), a vial of poison, a hammer, and a radioactive substance (RS). When the RS decays, the GC detects it and triggers the hammer that breaks the vial to release the poison, and the cat dies subsequently. Since radioactive decay is a random process, and there is no way to predict when it will happen, it is an either decayed or not decayed situation. In that case, an observer does not know whether the cat is alive or dead until he opens the box and observes the cat.

As per the Copenhagen interpretation, Schrödinger explained the state of the cat as both 'living and dead ... in equal parts', i.e., multiple likelihoods of a single event, until made an observation. I prefer to interpret the outcome of this thought experiment differently. When observed (associated with an observer wave function), the matter-wave would collapse into a single most probable state of classical observable out of many probabilities. It is the *quantum collapse,* or the *wave function collapse* due to the interaction between the two wave functions, one representing the matter states and the other representing the observer states. As scientists now believe that the universe is abounding with *consciousness wave,* any classical observable is the product of the *consciousness collapse.* Scientists have also started accepting that the consciousness wave is the gravitational wave predicted by Einstein.

The observer wave function travels through the Moana of universal consciousness, so, might get modified by interactions while traveling, which makes the observation unpredictable. It has nicely revealed through the famous dialogue between the two great personalities, Albert Einstein and Rabindranath Tagore, as appeared below.

Einstein: *Do you believe in the divine as isolated from the world?*

Tagore: *Not isolated. The infinite personality of man comprehends the universe. There cannot be anything that cannot be subsumed by the human personality, and this proves that the truth of the universe is human truth.*

I have taken a scientific fact to explain this—matter is composed of protons and electrons, with gaps between them; but matter may seem to be solid. Similarly, humanity is composed of individuals, yet they have their interconnection of human relationship, which gives living unity to man's world. The entire universe is linked up with us in a similar manner, it is a human universe.

I have pursued this thought through art, literature and the religious consciousness of man.

Einstein: *There are two different conceptions about the nature of the universe: (1) the world as a unity dependent on humanity, (2) the world as a reality independent of the human factor.*

Tagore: *When our universe is in harmony with Man, the eternal, we know it as truth, we feel it as beauty.*

Einstein: *This is the purely human conception of the universe.*

Tagore: *There can be no other conception. This world is a human world—the scientific view of it is also that of the scientific man. There is some standard of reason and enjoyment which gives it truth, the standard of the Eternal Man whose experiences are through our experiences.*

Einstein: *This is a realisation of the human entity.*

Tagore: *Yes, one eternal entity. We have to realise it through our emotions and activities. We realised the Supreme Man who has no individual limitations through our limitations. Science is concerned with that which is not confined to individuals; it is the impersonal human world of truths. Religion realises those truths and links them up with our deeper needs; our individual consciousness of truth gains universal significance. Religion applies values to truth, and we know this truth as good through our own harmony with it.*

Einstein: *Truth, then, or beauty is not independent of man?*

Tagore: *No.*

Einstein: *If there would be no human beings any more, the Apollo of Belvedere would no longer be beautiful.*

Tagore: *No.*

Einstein: *I agree with regard to this conception of Beauty, but not with regard to truth.*

Tagore: *Why not? Truth is realised through man.*

Einstein: *I cannot prove that my conception is right, but that is my religion.*

Tagore: *Beauty is in the ideal of perfect harmony which is in the universal being; truth the perfect comprehension of the Universal Mind. We individuals approach it through our own mistakes and blunders, through our accumulated experiences, through our illumined consciousness— how, otherwise, can we know truth?*

Einstein: *I cannot prove scientifically that truth must be conceived as a truth that is valid independent of humanity; but I believe it firmly. I believe, for instance, that the Pythagorean Theorem in geometry states something that is approximately true, independent of the existence of man. Anyway, if there is a reality independent of man, there is also a truth relative to this reality; and in the same way the negation of the first engenders a negation of the existence of the latter.*

Tagore: *Truth, which is one with the universal being, must essentially be human, otherwise whatever we individuals realise as true can never be called truth – at least the truth which is described as scientific and which only can be reached through the process of logic, in other words, by an organ of thoughts which is human. According to Indian philosophy there is Brahman, the absolute truth, which cannot be conceived by the isolation of the individual mind or described by words but can only be realised by completely merging the individual in its infinity. But such a truth cannot belong to Science. The nature of truth which we are discussing is an appearance – that is to say, what appears to be true to the human mind and therefore is human, and may be called maya or illusion.*

Einstein: *So according to your conception, which may be the Indian conception, it is not the illusion of the individual, but of humanity as a whole.*

Tagore: *The species also belongs to a unity, to humanity. Therefore, the entire human mind realises truth; the Indian or the European mind meet in a common realisation.*

Einstein: *The word species is used in German for all human beings, as a matter of fact; even the apes and the frogs would belong to it.*

Tagore: *In science we go through the discipline of eliminating the personal limitations of our individual minds and thus reach that comprehension of truth which is in the mind of the universal man.*

Einstein: *The problem begins whether truth is independent of our consciousness.*

Tagore: *What we call truth lies in the rational harmony between the subjective and objective aspects of reality, both of which belong to the super-personal man.*

Einstein: *Even in our everyday life we feel compelled to ascribe a reality independent of man to the objects we use. We do this to connect the experiences of our senses in a reasonable way. For instance, if nobody is in this house, yet that table remains where it is.*

Tagore: *Yes, it remains outside the individual mind, but not the universal mind. The table which I perceive is perceptible by the same kind of consciousness which I possess.*

Einstein: *If nobody would be in the house the table would exist all the same—but this is already illegitimate from your point of view—because we cannot explain what it means that the table is there, independently of us.*

Our natural point of view in regard to the existence of truth apart from humanity cannot be explained or proved, but it is a belief which nobody can lack—no primitive beings even. We attribute to truth a super-human objectivity; it is indispensable for us, this reality which is independent of our existence and our experience and our mind—though we cannot say what

it means.

Tagore: *Science has proved that the table as a solid object is an appearance and therefore that which the human mind perceives as a table would not exist if that mind were naught. At the same time, it must be admitted that the fact, that the ultimate physical reality is nothing but a multitude of separate revolving centres of electric force, also belongs to the human mind.*

In the apprehension of truth there is an eternal conflict between the universal human mind and the same mind confined in the individual. The perpetual process of reconciliation is being carried on in our science, philosophy, in our ethics. In any case, if there be any truth absolutely unrelated to humanity then for us it is absolutely non-existing.

It is not difficult to imagine a mind to which the sequence of things happens not in space but only in time like the sequence of notes in music. For such a mind such conception of reality is akin to the musical reality in which Pythagorean geometry can have no meaning. There is the reality of paper, infinitely different from the reality of literature. For the kind of mind possessed by the moth which eats that paper literature is absolutely non-existent, yet for man's mind literature has a greater value of truth than the paper itself. In a similar manner if there be some truth which has no sensuous or rational relation to the human mind, it will ever remain as nothing so long as we remain human beings.

Einstein: *Then I am more religious than you are!*

Tagore: *My religion is in the reconciliation of the Super-personal Man, the universal human spirit, in my own individual being.*

Consciousness also could be the unified field in the universe, which incorporates the gravitational wave. Physicist Amit Goswami has defined the quantum collapse as the downward causation of consciousness (2018),

"The quantum collapse is the possibility of the consciousness to choose from to precipitate an object; this is the downward causation of consciousness"

We know the mechanism of recognizing an object. The light from the body reaches the retina and stimulates its photoreceptors. Photoreceptors then send electrochemical signals to the brain. The brain interprets and recognizes that object. The reflection of light involves electromagnetic interaction between electrons of the atoms at the surface of the body and photons of the incident light.

Is there a quantum collapse of wave functions while seeing an object? Since an entity is constituted of atoms, seeing it involves the electromagnetic interaction. So, seeing a thing is a consequence of downward causation of consciousness. Thus, there is a quantum collapse. The processing of visual data takes place in the ventral visual stream, which is a hierarchical area in the brain that helps to recognise an object. But the processing of data requires few pieces of prior information such as knowledge about that object. For example, a chair has a typical structure that includes back support, four legs, and two arms. The article gets perceived through electrochemical signal processing in the brain. Those signals carry energy that interacts with our brain cells (due to energy-matter equivalence) and creates the perception of the object. However, in the absence of prior knowledge in the brain or missing data due to cell damage in a specific part of the brain, the recognition would fail.

How does the brain store data? Our brain has two types of memory areas, *short-term* and *long-term.* Short-term memory areas can store data for a few seconds to a minute. By repeating several times, data get transferred to the long-term memory location so that it can be used for a longer time, say, a few months to years. The brain stores data through the *coding process.* For an application purpose, such as recognition of an object, the stored data get decoded systematically. The same goes for a computer hard disk where the data get stored in ASCII (American Standard Code for Information Interchange) and decrypted into user language for a human user. Our brain is an energy-intensive organ that uses about 20 percent of the whole body's calorie requirement.

In conclusion, this chapter highlights the quantum concept of matter, energy-matter duality, energy-matter interaction, and consciousness. About the mass-energy equivalence, Einstein wrote:

"It followed from the special theory of relativity that mass and energy are both but different manifestations of the same thing—a somewhat unfamiliar conception for the average mind."

CHAPTER SIX

MIND-MATTER DUALITY

PURPOSE OF THIS CHAPTER: To know how the fundamental forces control our mind and its interaction with the neighbourhood.

"It is not enough to have a good mind; the main thing is to use it well."

— René Descartes

MIND, according to the Oxford dictionary, is the element of human being that enables them to be aware of the world and their experiences, to think, and to feel; the faculty of consciousness and thought. MATTER, on the other hand, is a physical substance and is distinct from mind and spirit; it occupies space and has mass. The matter could be either animate or inanimate, microscopic, or mammoth. The intellect without soma is non-existent, but the reverse is not always true. I shall first discuss *mind-body duality* and then *mind-matter duality*, matter excluding the human body.

The human being has a mind as well as a body, which

contains a brain. Collectively they bring a duality problem, mind-body duality, which has originated from a statement of René Descartes, a 17th-century French philosopher mathematician:

"I think; therefore I am."

The mind is a non-materialistic, non-extended substance that involves various activities, as mentioned above. Matter, on the other hand, obeys the laws of physics. Though the human body is a matter, it does not follow the laws of physics. The reason behind, as Descartes believed, it is causally affected by the mind, which relevantly produces certain mental events mediated by the pineal gland in the brain. For example, when we wish to walk, we can do so, i.e., mind influences the body; when we get hit by a hard object, our mind feels pain, i.e., the body influences mind. Thus, the mind and the body (i.e., the brain) are two different objects, which are interlinked. Descartes's mind-body dualistic theory, known as *interactionism*, raised a vital question, *how is that causal interaction possible?*

LEIBNIZ VIEW: According to German mathematician and philosopher Gottfried Leibniz, the mind never interacts with the brain (i.e., body); they function in parallel. It is called *parallelism* between mind and brain, which is mediated by consciousness. It is a phenomenological view of the causal interaction between the mind and the body, where the linker is consciousness.

SALT VIEW: Dr. William Salt, MD, from Columbus, OH, explained that the brain plays a pivotal role in translating the content of the mind into nerve cell firing and chemical release, i.e., into *neurosignatures,* which intimately affect the physiology and biochemistry of the body. It is a materialistic view of the mind-body interaction, where the brain interacts directly with the mind, and consciousness is not needed.

BERKLEY VIEW: Irish philosopher Gorge Berkeley claimed, the body is merely a perception of the mind; it is real, and the body is unreal.

BEHAVIOURISTS AND BIOLOGISTS VIEW: Behaviourists, in contrast, disregard the mind, as it is beyond any scientific and objective studies; some behaviourists even do not recognise the existence of the mind at all. Biologists also suggest ignoring it due to a lack of physical form. Thus, both behaviourists and biologists believe in real materials that can be seen, felt, and touched. It is *materialism,* which reveals that the mind is the brain. It is in parity with Dr. Salt's view.

PSYCHOLOGISTS VIEW: Psychologists interpret the human brain as the CPU of a computer and the mind as the software. The difference is that software processes digit where mind processes meaning. Though it is practically non-observable, the existence of the mind can is verifiable through experience.

LET US ANALYSE: Supposing mind-brain duality exists *when the brain behaves as the mind or the mind like the brain?* According to the wave-particle duality theorem, particles can act as a wave as well. For example, the Young double-slit diffraction experiment confirms the wave behaviour, and the photoelectric effect proves the particle behaviour of light. Light is electromagnetic radiation, which is a manifestation of energy. So,

energy also can be expressed in both forms:

$$E = mc^2 = mc.c = pc = \frac{hc}{\lambda} \qquad (6.1)$$

Thus, the momentum (*mc*) is $p = E/c$, and the wavelength is $\lambda = hc/E$ of a particle of mass *m* propagating at speed *c*. Eq. 6.1 suggests that energy has relations with both p and λ. In other words, energy mediates the wave-particle duality; the particle behaviour dominates when its speed is low, else the wave behaviour dominates.

For brain and mind, is there a mediator? Yes…consciousness is there, according to the German philosopher Gottfried Leibniz, considering the brain is the matter, and the mind is the wave. However, both energy and consciousness are non-measurable entities. We can realise energy only in certain forms like heat or light. Similarly, consciousness is realised while seeing an object or creating a painting, and so on.

According to quantum mechanics, the matter is the wave, and the material-wave is consciousness. It has engrossed in Buddhist philosophy as well. Seeing an object is eloquent to the consciousness collapse in the brain that gets recognised through neural signal processing. It is the downward causation of consciousness (Goswami, 2018) that makes a relation with the cerebrum.

When does consciousness relate to the mind? When we imagine something. Let us revisit the Tagore's version, "Science has proved that the table as a solid object is an appearance and therefore that which the human mind perceives as a table would not exist if that mind were naught. At the same time, it must be admitted that the ultimate physical reality is nothing but a multitude of separate revolving centres of electric force, also belongs to the human mind." This human mind is an individualised form of the universal mind, as Tagore explained. Every object belongs to that universal mind, which is the consciousness, and it collapses into the human brain while visualising an object, event, or scenery.

Can the mind work without the brain? According to the neurologist Jacob Sage, MD:

"I am a working neurologist who sees brain disease causing mental dysfunction every day…There is a famous case of Phineas Gage, however, in which brain damage to the frontal lobes of the brain by a railfoad spike turned a sober, hardworking man into a lout. His mind was altered because his brain was altered. He was different person after that spike went through his brain."

He also said,

"…nothing more than the ability of the brain to acquire information and all the content that the information contains and the ability to get all that information into and out of memory……brain can do all that, and, therefore, a functioning brain is identical to a conscious mind."

It is the same that Dr. Salt did express. Professor William Klemm, a senior professor of Neuroscience in Texas, expressed:

"Brain has conspicuous functional states, ranging from intense conscious concentration to drowsiness, to sleep, to coma, to death. Neural electrical activity correlates in a systematic way with those

state changes."

Using an analogy, I would say that the mind is like the light from an electric bulb. It is generated only by the application of an action potential (see Chapter Three) across a neural circuit in the brain as an electrical potential across the electrical circuit for light. Like light, the mind also has colours. It is a product of the brain, as luminescence is a product of the bulb. A switch triggers the potential across a circuit. In the case of neural current, consciousness ignites the action potential across the circuitry of neurons. In Hindi, "dimug ka batti jalao" means switch on the light of your brain. The neural flow creates an impulse pattern that is responsible for many functions of the brain, such as thinking and imagining in the presence of a conscious mind. Thus, there is no mind-brain or mind-body duality. The brain generates the mind that can imagine any kind of fantasy, like the Aladdin's magic lamp jinn.

In the mind-matter duality, the matter could be either living, except the human body, or nonliving object. In the previous chapter, I have discussed energy-matter duality. It is the same as wave-matter duality; because energy can be either light or heat waves. If the mind is a wave, mind-matter duality would also be like wave-matter duality. As the states of the brain are the mind and, at the same time, the brain is a matter, there is no mind-matter duality as such. Rather the mind-matter relationship or the mind-matter interaction is relevant as it helps us to make a decision, makes us happy, activates our creativity, and so on. In this regard, it is worth mentioning a comment by the American comedian Jack Benny,

"If you don't mind, it doesn't matter."

Let me re-express above comment differently, like 'if you do not involve the mind, matter does not appear', that is the mind perceives matter, as claimed by Gorge Berkeley.

What mediates the interaction between mind and matter? The cerebral cortex or cerebrum of the brain controls our ability in reading, thinking, learning, speech, emotions, planned muscle movements (like walking), vision, hearing, and other senses. It is a part of our brain where the mind originates, but the mind is not a matter or any somatic fragment of the brain.

How does the mind originate from the brain? The Greek physician Hippocrates stated (Francis Adams, 1886):

"And men ought to know that from nothing else but thence [from the brain] come joys, delights, laughter and sports, and sorrows, grief, despondency, and lamentations. And by this, in an especial manner, we acquire wisdom and knowledge, and see and hear, and know what are foul and what are fair, what are bad and what are good, what are sweet, and what unsavoury... And by the same organ we become mad and delirious, and fears and terrors assail us... All those things we endure from the brain, when it is not healthy... In those ways I am of the opinion that the brain exercises the greatest power in the man. This is the interpreter to us of those things which emanate from the air, when it [the brain] happens to be in a sound state."

What did he mean by *air*? Consciousness! What is consciousness? According to the Oxford Dictionary, it is the state of being aware of and responsive to the surroundings. Tagore

explained that the human mind, which gets generated from the brain, is a localised form of consciousness. According to the Chinese Philosopher Lao Tzu,

"The key to growth is the production of higher dimensions of consciousness into our awareness."

So, a matter (i.e., an object) is a part of the consciousness, and it gets observed through the consciousness collapse. When we see an entity, it gets recognised through neural signaling processes in the brain. A synchronised combination of electrochemical signals that are created by the synchronised firing of the synapses of neurons is called *brainwave*, which gives a sign of a specific state of consciousness, thought, and mood. Brain waves are studied using an electroencephalogram (EEG) instrument. It measures the oscillation of the electrochemical current in the brain through the electrodes placed at different positions on the scalp.

German physiologist and psychiatrist Hans Berger invented the EEG technique and measured the first human EEG. Brain waves those create spectra of the human mind have five different frequency bands (Bergland, 2015): *delta wave* (δ-wave) in the frequency band of 0.5–3 Hz, *theta wave* (θ-wave) in 3–8 Hz, *alpha wave* (α-wave) in 8–12 Hz, *beta wave* (β-wave) in 12–30 Hz, and *gamma wave* (γ-wave) in 30–100 Hz. Representations of those waves are as shown in Figure 6.1. The low-frequency delta and theta waves are active in the sub-conscious states of the brain, such as deep sleep. Alpha waves during daydreaming or consciously practicing mindfulness or meditation. Beta waves dominate our normal waking states (i.e., alert, attentive, and focused) of mind when attention is towards cognitive and other tasks such as problem-solving or decision-making activities. Gamma waves (usually hover around 40 Hz) are associated with the simultaneous processing of information from different areas of the brain and other higher states of conscious perception. Therefore, the brain is always active in analyzing and controlling different states of the mind. However, the mental states actively collaborate with the brain only when consciousness boosts it. Thus, consciousness alertness is the driving force of the mind, and electrochemical processes are the driving force of the brain. Table 6.1 provides a glance at various brain waves connected with different mental states.

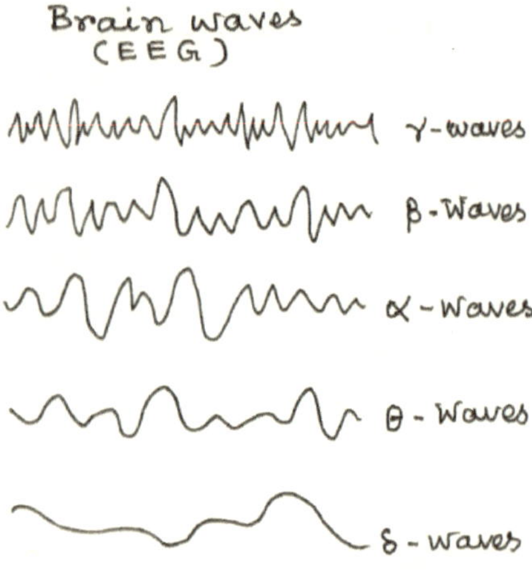

Figure 6.1. Five brain waves in different frequency bands, given at the right ends.

Table 6.1. Various brain waves connected with different mental states.

Brain waves (frequency range in Hz)	Mental state
Delta (0.5 – 3)	Deep dreamless sleep, unconscious
Theta (3 – 8)	Light sleep with REM, deep meditation, creative, intuitive, fantasy, memory
Alpha (8 – 12)	Relaxed but conscious, light meditation, super learning, day dreaming, creative
Beta (12 – 30)	Normal awake state, concentration, alert, attentive, focused, problem solving, decision making
Gamma (30 – 100) (Usually hover around 40)	Simultaneous processing of information from different areas of the brain, motor functions, higher states of consciousness, insight, peak focus, perception

The Neurochemicals of happiness, an article published in Psychology Today (Bergland, 2012), has reported about the seven brain molecules, as described below, which make us feel great.

(i) *Endocannabinoids*—Cannabinoid is a product of cannabis (a genus of flowering plants in the family of Cannabaceae). The most well-known cannabinoid is the delta-9-tetrahydrocannabinol (Δ9-THC; $C_{21}H_{30}O_2$), which is the prime psychoactive ingredient in cannabis. Endocannabinoids (known as 'bliss molecule') are self-produced lipid messengers that work on cannabinoid receptors (CB-1 and CB-2), which are part of the endocannabinoid system found in cells that alter neurotransmitter release in the brain, and, consequently, alters the perceptions and states of mind in various ways. A study at the University of Arizona indicates that humans and dogs show significantly increased cannabinoids following sustained running.

(ii) *Dopamine*—Dopamine (known as 'reward molecule'; $C_8H_{11}NO_2$) is a neurotransmitter produced naturally in the human body, which sends signals from the body to the brain. It is responsible for reward-driven behaviour and pleasure-seeking. Dopamine plays a significant role in controlling a person's movements as well as his emotional responses. A right balance of dopamine is, thus, paramount for both physical and mental well-being. There is evidence that people with extraverted personalities have higher levels of dopamine than introverted people.

(iii) *Oxytocin*—Oxytocin (known as 'bonding molecule'; $C_{43}H_{66}N_{12}O_{12}S_2$) is a peptide hormone and neuropeptide produced in the hypothalamus and released by the posterior pituitary, and works for human bonding, trust, and loyalty. The report says that strong emotional bonding between humans and dogs may have a biological basis of oxytocin.

(iv) *Endorphin*—Endorphin (known as 'pain-killing molecule') is a hormone that is secreted by the pituitary gland and hypothalamus in the brain and nervous system during strenuous physical exertion. Endorphins (i.e., self-produced morphine) activate the body's opiate receptors, causing an analgesic effect.

(v) *GABA*—γ-aminobutyric acid or GABA ($C_4H_9NO_2$, known as an 'anti-anxiety molecule') is the chief inhibitory neurotransmitter in developed mammalians' central nervous system (CNS) that slows down the firing of neurons and creates a sense of calmness. Yoga practicing causes more GABA production.

(vi) *Serotonin*—Serotonin or 5-hydroxytryptamine ($C_{10}H_{12}N_2O$, known as the 'confidence molecule') is a monoamine neurotransmitter. It is also known as the 'happy molecule' because it contributes to well-being and happiness. It serves a wide variety of functions in our bodies. Taking challenges regularly and pursuing things that reinforce a sense of purpose, meaning, and accomplishment do increase serotonin production. Any successful performance does produce a feedback loop that reinforces behaviours that build self-esteem, reduce insecurity, and create an upward spiral of more and more serotonin.

(vii) *Adrenaline*—Adrenaline (also known as the 'energy molecule'; $C_9H_{13}NO_3$), technically known as epinephrine, plays a vital role in the fight-or-flight mechanism. Adrenaline gets produced by both adrenal glands and a small number of neurons in the medulla oblongata, where it acts as a neurotransmitter in regulating the visceral functions. The release of adrenaline is exhilarating, and it creates a surge in energy that makes a person feel alive. It can be an antidote for boredom, malaise, and stagnation. By taking short rapid breaths and contracting muscles, the adrenaline rush can get built up in the body.

Those seven molecules and few other chemicals assist in electrochemical signalling through neural networks in our brain as well as other parts of our body. The action potentials originate through the firing of neurotransmitters at the synapses. Consequently, electrochemical signals get produced, which travel through the neural pathways. The passage of communication realises through sequential electrostatic interactions along the cell wall from soma to the axon terminal. Subsequently, relevant areas in the brain accomplish various activities like recognition of shape, colour, smell, light, thinking and decision making, body movements, and leisure under the conscious state of mind as well as dreaming and breathing under the sub-conscious state of mind. Those brain activities make us aware of our neighbourhood and the rest of the world and help in developing love and affection with other persons, and passion for arts and music.

In THE BEGINNING, I mentioned some relationships such as *friendship*, *parent-child relationship*, and *the teacher-student relationship.* These relationships modify with time. Having a relation, or even a passion for something, is a non-materialistic event, though it involves the brain. *How do those relationships, or mental ardent, depend on the neurochemical signals, i.e., brain waves?*

Consider friendship. *How does it grow up between two parsons?* It starts with awareness about each other, e.g., knowing personal details or exchanging thoughts on a common subject. It may happen in meeting personally or talking over the telephone or writing via social media. One may enquire about the place of birth, family and educational background, and hobby of the other

person. Does it help in making friendships? Yes, it does because human beings are homophilic (Alonsoin, 2018). Our benevolence grows based on the resemblance in a wide range of factors such as age, race, religion, socioeconomic status, academic level, political tendency, assessment of cleanliness and hygiene, even strength of the handshake. Closeness at the genetic level is also very crucial in making friends. But we shall avoid this topic here.

Carolyn Parkinson and co-researchers of the Department of Psychology, University of California, Los Angeles, investigated the neurological response patterns of a group of students while freely viewing naturalistic movies. Researchers scanned the pupils' brains using the functional MRI (fMRI) technique. They intended to get the answer of, do similarities among friends reflect profound connection in how they perceive, interpret, and respond to the events? The outcome indeed confirmed the neural homophilicity among those friends, and that the neural activities are exceptionally matching among friends. This affinity decreases with increasing distance in a real-world social network (Parkinson, 2018).

Friendship depends on the coherency in brain activities among friends, which depend on how they interact with their surroundings. Consciousness, intertwined with the mind or brain waves, does play a crucial role in that interaction. Modern science believes that the unified field (of four fundamental forces) of the universe is the consciousness field or the *air* stated by Hippocrates. As dropping a stone in a pond creates a wave, stimulation in the consciousness field also creates consciousness waves. Those sway the brain to trigger action potentials through neurons, i.e., chemical signals and, in turn, brain waves. The superposition of the two waves causes a localised consciousness-collapse. It creates an image or a sensation of the stimulation that had propagated through the air. In their study, Parkinson and co-researchers mapped the brain waves of students, generated by the consciousness waves in the air due to the stimuli from the movie show.

The consciousness waves originate from different types of stimuli, e.g., sound or light. The frequency of a brain wave, due to identical consciousness waves, varies from person to person, and that provides knowledge about different mental conditions among them. The brain wave frequencies may also vary depending on other factors that include the structure of the neural network in the brain and past experiences. Such aspects individualise one person from the other. Matching of brain waves among a group of adult human beings is relatively less probable compared to a group of children or youngsters as experience does affect a friendship.

The parent-child relationship is the most complex one among the three considered here. It reveals a unique bonding between the child and the parents, and it is almost worldly alike. When a baby grows up from a school-going child to an adult, this relationship automatically modifies based on certain expectations in day-to-day life. It starts with unconditional love from parents, irrespective of the behaviour or achievements of the child. It is a need for the child and not for the parents (Sunderland, 2006). Parents love their babies passionately. However, the baby needs to sense the love that is crucial for his or her brain development. This sense arises over time through interactions during several

posturings, such as hugging, rocking, and holding (Perry, 2004). How does this sense of love and attachment grow in the minds of both the baby and the parents? Scientists found, while the baby and the mother are in eye contact, the brainwaves from both get synchronised, i.e., matching the frequencies of chemical impulses in their brains. This synchronisation of brain waves creates a sense of mutual attraction between a mother and her baby.

The parent-child relationship also depends on many other factors. The baby's heartbeat, or even emotions, synchronises while interacting with her primary caregiver (Feldman, 2011). Those neural and physiological synchronisations together facilitate a parent-child unique bond. Thus, frequent interactions during infancy have a positive impact on a child-parent attachment, as well as on brain development and functions for the rest of a person's life. A lack of such closeness during childhood can affect the relationships and the ability of the person to communicate effectively with others in his adult life. Experiences from those initial attachments get encoded in the neural circuitry of the brain by 12–18 months of age, entirely in implicit memory without awareness. Those initial practices become the rules in the lifelong relation.

The bond becomes multifaceted when the baby grows up and starts going to school. During the pre-school stage, love remains unconditional, and the baby can sense the love and attention towards her. As the baby's activity with the outer world increases, the mother enjoys that and tries to teach many things such as mannerism, interactions with friends and teachers, the meaning of words, and the construction of sentences. Slowly and unknowingly, the mother develops an expectation of quick learning and advancement from her child. Nothing wrong in hoping from her side, but it becomes a *condition* for the child to be loved. Thus, love turns conditional, which a child usually dislikes. He or she starts protesting against such nick, which unknowingly gives a new hank in their relationship. Again, there is nothing wrong with the parent's side to be concerned about the child.

At a young age, a child gets emotionally attached to some friends who can regulate his or her mind significantly. Slowly, the frequencies of brain waves between the parents and the child start deviating. This process of transformation of the parent-child relationship from childhood to adulthood continues, which almost alike through the countries, races, and religions. The initial parent-child bonding may or may not remain intact. Sometimes that may remain unaltered from the parent-side towards their descendant, but usually not in the other way (with few exceptions). The mind, which is a composition of several brain waves, plays a significant role in this relationship. So, the neural signals in the brain are instrumental in a parent-child relationship, as in a friendship.

The teacher-student relationship has a societal impact. The practices of this relationship fall between the friendship and the child-parent relationship. A good teacher-student relationship demands a teacher to be like a friend as well as a parent. A supportive relationship boosts the achievement of a tutee; it also protects him from mental stress. Unfortunately, many times love from a teacher towards a student varies with his conduct in different activities, which is a *condition*. It fabricates mental discriminations among pupils, which prolongs till the end of school life, i.e., when students are at their teen ages. For a successful teacher, some criteria need to fulfill. According to Mary Helen Immordino-Yang, a cognitive neuroscientist at the University of Southern California,

"A lot of teachers ... have really strong abilities to engage socially with the students, but then it's not enough. You have to go much deeper than that and actually start to engage with students around their curiosity, their interests, their habits of mind through understanding and approaching material to really be an effective teacher."

Vicki Nishioka, a senior researcher in Education Northwest who studies teacher-student relationships said,

"Sometimes teachers don't understand the importance that their relationship with each student has on that student's identity and sense of belonging. What gets in the way of that is a more authoritarian kind of discipline and interaction approach with students, which really doesn't work."

How to make this relationship more successful? The most probable answer is through *empathy*. What is empathy? According to the Cambridge dictionary, it is the ability of a person to share feelings or experiences of another person by imagining what it would be in his situation. Empathy has two aspects, cognitive and emotional, which are the kind of reactions of an individual to the shared experiences of another. It creates a matching of the minds of both persons, which causes the secretion of appropriate hormones in the brain and reduces the feeling of guilt, distress, or pain from the deceived mind. Recent studies revealed that a teacher who cultivates empathy for and with his students could manage their behaviour and academic engagement in better ways.

Besides developing relationships, mind-matter interaction plays many vital roles in human life; for example, following a hobby such as listening to music, reading books, traveling, singing, painting, and many others. A hobby should be such that it helps relaxing mental and physical stresses, increase performing efficiency and creativity of a person. It has an immense effect on both the brain and the heart.

MUSIC AGAIN! When melodies played on a musical instrument such as guitar or harmonica or saxophone, that symphony reaches our ears through the air and creates vibrations in the eardrum. Those vibrations stimulate action potentials at the neuron terminals and generate chemical signals and brain waves with specific frequencies. It is well-proven that listening to music increases the level of the dopamine in the cerebrum. Dopamine is known as a *reward molecule* or *motivational molecule* and an integral part of the pleasure-reward system of the brain. Being a passionate harmonica player, I do feel immense pleasure while playing melodies on harmonica and relax in both mind and body.

Music can shape our emotions. Some songs can make us feel happy, while others can even bring tears to our eyes. I discern content in my heart while playing or listening to songs of Tagore (*Rabindra sangeet* in Bengali). Both the meaning and the melody of those songs create heavenly feelings in my ego. *Why do I feel so much pleasure in Rabindra sangeet*? Because those songs are rooted in a slow melody that stimulates low-frequency waves in the air, which activate either theta or alpha waves in my brain. For some time, my mind goes into the meditation mode and connects with the divine melody of nature. Such a state of mind gets accompanied by the flow of dopamine that generates motivation in me.

What type of music does, in general, a person like? Some prefer slow music, while others fast music. The taste of music depends on the mental growth of a person. For example, I am inferior in interacting with people and usually prefer to stay alone; accordingly, my brain has gathered experience of life. On the other hand, an interactive person generally enjoys fast melodies. Thus, the same jingle may not create the same feeling in every mind. Even so, music has a universal effect. Researchers from McGill University, University of Montreal, and Technische Universität Berlin reported in Frontiers in Psychology about worldly effects of music worldwide regardless of cultural diversities. Listening to and playing music can make a person smarter, happier, healthier, and more productive at all stages of life.

Can music help in healing mental problems? According to Dr. Maoshing Ni, a doctor of Chinese medicine and leading authority on Taoist anti-aging medicine, and author of the books, *Secrets of Longevity, Secrets of Self-Healing, Second Spring: Hundreds of Natural Secrets for Women, Secrets of Longevity 8-Week Program,* and *The Natural Health Dictionary,* for thousands of years music is in use in medicine. Ancient Greek philosophers believed that music has a healing effect on the body and the soul.

Singing and chanting have been a part of Native American healing ceremonies for millennia. In the Ottoman Empire, music was in use as a therapy for mental illness. A more formal approach to music therapy began after World War II when researchers observed that music had a positive effect on emotionally disturbed veterans. Music therapy can reduce high blood pressure, depression, and sleeplessness. People who are high in one of the five personality dimensions called *openness to experience* are likely to experience the most chilled while listening to music (Nusbaum and Silvia, 2010). A review article in Neurochemistry revealed that music could boost the function of the immune system in the body and lessen the stress level. The lead researcher, Joke Bradt of the Department of Creative Arts Therapies at Drexel University in Philadelphia, explained:

"The evidence suggests that music interventions may be useful as a complementary treatment to people with cancer."

One of the most remarkable successes in music therapy has on Alzheimer patients. Advanced patients lose their ability to interacting with others and eventually stop speaking altogether. Investigation showed that listening to familiar music, often patients as though turn on and start singing along. The effect arises probably due to a significant increase in the melatonin concentration in serum after music therapy, which may have contributed to a relaxed and calm mood of the patients (Kumar et al., 1999).

In general, any form of art, such as painting, music, and dance, can improve the functions of our brains. According to Renee Phillips, founder Director and Curator of Manhattan Arts International, scientific evidence confirmed arts enhance brain activity. She wrote this in an article in *The Healing Power of Arts & Science & Art*. Art influences the brain wave patterns and emotions, the nervous system, and helps to raise the serotonin level to increase the happiness and confidence of a person.

How does painting, a visual art, sway our brain? If a

painted picture is shown to a group of people and asked to give their comments, you would receive mixed responses. For example, some will comment that it was nice, but did not know why! Some others will say that it was ok. However, some of them will comment critically! If you analyze those responses, you would realize that the first two classes of people are not so keen about paintings. But people in the first category at least enjoy some features of paint, such as the colour combination that soothes their mind. On the other hand, people in the third camp have a good perception of colors and fine arts. Different paintings may have different levels of influence on an individual mind depending on his character that encompasses genetic makeup, cultural annals, and experiences.

Which part of our brain is responsible for the appraisal of a painting? The right hemisphere administers tasks related to creativity and the arts. It is the aesthetic system, which cherishes any form of art, such as painting. According to Brown and Gao (2011), the *anterior insula* in the brain is responsible for the aesthetic appraisal of an art form; it sits within one of the deep folds of the cerebral cortex. Painting helps in communicating between the minds of the artist and a viewer. A real critic would be able to interpret the thought of the painter through the paint.

The concept of painting was generated in the mind of early human beings when there was no script. The drawing was one of the ways of communication for them. Today, paintings communicate many powerful messages (e.g., for peace, for protest) from the minds of painters to viewers. Neurochemical signals are produced in the brain of the artist while creating a painting, and those signals stimulate the consciousness waves in the air. A real critic can feel the signatures of those waves produced in the artistic brain, as the painting does reflect that. Similar waves are created in the intellectual mind of the interpreter while observing that painting. It is like exchanging thoughts on a one-to-one basis through the picture as the medium. Many factors may impact a viewer's mind, as mentioned above. Because of the synchronisation of the brain waves in the two brains, a critic will be able to interpret the message in the painting. Some of them buy those artworks and keep in their living rooms to sustain that fondness in their minds.

Mind-matter interaction does govern a person while reading a novel or a story. Reading is a hobby, which many people have, and a good writer knows how to exploit it. In a storyline, the writer draws a picture of life using their style of communication with the readers, which may be emotional or social, or other types. It is the skill of a writer to dominate the minds of the readers. Sometimes, it creates excitement or a sense of pleasure, depending on the subject of the novel. Personal interest in a piece of writing depends on many factors such as mental growth, educational and professional accounts, and experiences in life. Writers who are capable of impelling the readers through their novels, stories, or poems become very popular and even idol to a community. For example, J. K. Rowling became extremely popular with teenagers throughout the world for her children series of *Harry Potter*. Hollywood movies have been produced in those imaginary tales. Music, novels, or films can create long-lasting impacts on society. Thus, arts can be put-to-use to motivate the minds of students, teachers, professionals, business personalities, and politicians; after all, art is a game of intellect.

Thus, mind-matter interaction plays a very crucial role in our lives as well as society. It involves chemical signal processing in the brain that constitutes the mind of a person and consciousness of the universe. A robot is not a human being; so, it cannot be a part of a society where every living being has a philosophical mind. Professor Albert Einstein said:

"I fear the day that technology will surpass our human interaction. The world will have a generation of idiots."

CONCLUSIONS

PURPOSE AT THE END: Reverberation of this entire expedition and response to the fundamental question, *what is the identity of God?*

This journey had started with a fundamental question on the identification of God of this universe. As the expedition and exploration have ended, I need to conclude. With acquired results and indication, I can say that the voyage was momentous. The reverberation of the journey persists in my mind, and I would like to look back once again through the entire tour very briefly, before writing the conclusion.

The beginning was on the history of the universe created through a massive boom, called the Big Bang. Within a fraction of second, i.e., 10^{-43} second (known as Planck time), the universe had experienced a rapid expansion, called inflation. After that (10^{-36} seconds onwards), a steady escalation started that is continuing till now. Before the Big Bang, the giant universe had confined into a tiny room of the darkness of the space-time regime.

What was going on within that confinement? Probably, there was a tremendous fight between two opposing forces, one against the other to its integrity. Which one was good and which one was evil that is irrelevant now, as all that happened were in favor of life.

Being intelligent creatures of this planet, we must analyse the cause behind every event with scientific cognizance. Before the Big Bang, as the giant cosmos was concealing into a very tiny space, so it was almost infinitely dense. A strong repulsive force will always try to break such a highly condensed body. How could then it remained stable till the Big Band? A hefty attractive pull in the heart of that tiny giant probably fought against the strong repulsive force. But failed! Though the drawing effort was for integrity, it was the repellent force that gave the birth of this universe. What was that force, and does it exist now? Though the journey has ended, these questions remained unexplored.

Is this universe entangled with the web of consciousness since its birth? Scientists do firmly believe it now. They also suspect that consciousness is the unified expression of all fundamental forces. Every object in this universe, either with or without life, including the human body, procures those fundamental forces. In other words, no matter could originate in this universe without those fundamental forces. Buddhist philosophy also trusts on the consciousness web. The great personalities like Albert Einstein and Rabindranath Tagore were mesmerized by the universal consciousness. Einstein said:

"Brief is this existence, as a visit in a strange house. The path to be pursued is poorly lit by a flickering consciousness."

He also said:

"Our task must be to free ourselves by widening our circle of compassion to embrace all living creatures and the whole of nature and its beauty."

Tagore wrote in a song:

> The sky, adorned with stars and sun
> The world, full of so many lives
> Within this glorious vast expanse
> Here I am, with a place to be, a role to play
> I am amazed and wondering,
> I see my song find itself here…
> Along with the eternal flow of the time
> The way this world sways
> Through high and low tide…
> I feel the very same rhythm
> In the flow of blood in my veins…
> I am amazed and wondering,
> I see my song find itself here…
> I walk on this earth, barefoot on the grass
> Surprised and adored by the aroma of the flowers
> All I see seems made out of bliss…
> I am amazed and wondering,
> I see my song find itself here…
> My eyes are wide open, ears are eager
> My life is pouring out like a flowing river
> My heart seeks out for the unknown
> That lies within everything known…
> I am amazed and wondering,
> I see my song find itself here…

Every human being got empowered by self-consciousness, which has entangled with the global consciousness web. If we could recognise that connection, we could understand the language of nature, the language of the world, and the language of the universe. If we could learn the language of the world, everything in this world would make sense to us, as the shepherd boy thought (Coelho, 2018):

"I am learning the Language of the World, and everything in the world is beginning to make sense to me…even the flight of the hawks."

As the universe had expanded and cooled further, the elementary particles such as an electron, positron, quark, anti-quark, and gluon got manufactured from the nascent spirit of the Big Bang. That event advocated for the energy-mass equivalence relationship. The universe was so hot initially that the fundamental forces got overshadowed by the repelling impetus of high thermal energy. With considerable cooling of the cosmos, attraction due to the nuclear and the electromagnetic forces came into play, which propelled the formation of the nuclei and then atoms from the elementary particles. Afterward, the gravitational attraction also became prominent, which pulled gas particles and dust into clusters of giant structures such as planets and stars.

From where have those attractive forces emerged? Most likely, those have popped up from the pre-Big Bang infinitesimal seed. Immediately after the Big Bang, those forces got blended into one. With the expansion and cooling down of the universe, those forces evolved with different manifestations, which became known as the four fundamental forces, i.e., strong and weak nuclear forces, electromagnetic and gravitational forces. Those fundamental forces have created the universe, galaxies, stars, and all planets such as our Earth. Those forces help our Earth to evolve perpetually with the changing environment. That is how the Earth has transformed from a vulnerable to a stable blue living planet. Without those forces, life would not have originated. So, every matter in this universe has the blessings of fundamental forces. Scientists now think that those fundamental forces are the ingredients of an integrated one. And that might be eternal God, both before as well as after the Big Bang.

Through evolutions, life on Earth has restructured from unicellular organisms to multicellular animals and plants; oxygen-deficient to oxygen-supported metabolites; amoeba to modern human beings. Those transformations could happen due to the prolonged evolution of matter upon interactions with the environment, which could not have occurred without involving the fundamental forces. Those forces exist everywhere, supervise our body and mind, our health, and the interaction of medicine with the human physiology.

The nanoscience and nanotechnology, in the modern world, are blessed by those fundamental forces. Scientists are searching for alternative methods in medicines, such as targeted drug delivery, cancer therapy, and improved diagnosis of diseases by generating high contrast images of body organs using nanomaterials and nanotechnology. The alternate nanomaterials-based methods in medicine are called *nanomedicine*. Successful clinical tests of nanomedicine would solve issues of side effects of conventional drugs and assure healthier lives on Earth. But with

the modernization of existence, we are destroying our natural ecosystem, which may lead to untoward interactions between matter and the environment. A situation may come when no advanced medicine can keep us safe. We should be alert!

The universe has two ingredients: matter and consciousness. One is materialistic, and the other non-materialistic. There is a firm intermingling between the two. Consciousness has mingled with every material, starting from the elementary particles to the highest intellectual animals. In our brain, consciousness gets quenched as the mind, which helps us to see, think, imagine, or solve problems. Great personalities, Einstein and Tagore, could connect their mind with the universal consciousness.

Are consciousness and united force the same?

It is still an open question.

In 1925, Einstein shared his concern about the creation of the universe with one of his young students, Esther Salaman, with the following statement:

"I want to know how God created this world. I'm not interested in this or that phenomenon, in the spectrum of this or that element. I want to know His thoughts; the rest are just details."

His thoughts must be unity and love. Mahatma Gandhi also said:

"Let us work together for unity and love"

Now the final question: *What is the identity of GOD?* It is the UNIFIED FORCE. *How HE created the universe?* HE conducted some interactions to unify all matter in this universe. Unified force and consciousness are the two sides of God. The goal of the modern physics is to establish the existence of the united force and the *theory of everything* (TOE), which would answer all questions involving materials to mind. Einstein once said,

"I do not believe in a personal God and I have never denied this but have expressed it clearly. If something is in me which can be called religious then it is the unbounded admiration for the structure of the world so far as our science can reveal it."

Tagore said,

"Those who think to reach God by running away from the world, when and where do they expect to meet him? We are reaching him here in this very spot, now at this moment."

Is there any difference between the statements of the two great men? No. God is the unified force, the consciousness. God's actions are the interactions that have created this stunning nature around us, and our creative conscious minds. So, we must communicate with our Mother Nature. Before ending, let us sing together the following song of Tagore:

> *How well you sing, o accomplished artist!*
>
> *Stunned, I merely listen.*
>
> *The light of your melodies spreads across the world;*
>
> *The breeze of your notes flits through the sky,*
>
> *Splintered, the stones speed anxiously*
>
> *Flowing away in the stream of melodies.*

I feel like singing in the same style,

But cannot capture the notes.

What I Want to say, I cannot.

Defeated, my soul cries,

Weaving a web of notes all around.

What snares you have lured me into!

BIBLIOGRAPHY

Adams F, *The Genuine Works of Hippocrates*; *Volume 1*, William Wood and Company, New York (1886); ISBN: 9781362591290.

Alonsoin JR in *Neuroscience of friendship*, December 10 (2018); https://mappingignorance.org/2018/12/10/neuroscience-of-friendship.

Bergland C, The Neurochemicals of Happiness, *Psychology Today*, November 29 (2012).

Bergland C, Alpha Brain Waves Boost Creativity and Reduce Depression, *Psychology Today*, April 17 (2015).

Bergland C, The Neuroscience of Making a Decision, *Psychology Today*, May 06 (2015).

Bloch F, Zur Theorie des Ferromagnetismus, *Zeitschrift für Physik* 61 (1930) 206.

Bonner JT, *First signals. The evolution of multicellular development*, Princeton University Press, Princeton, NJ (2000).

Brown S, Gao X, The neuroscience of beauty: How does the brain appreciate art? *Scientific American*, September 27 (2011).

Burbidge G, Sir Fred Hoyle. 24 June 1915–20 August 2001 Elected FRS 1957, *Biographical Memoirs of Fellows of the Royal Society* 49(2003) 213; DOI:10.1098/rsbm.2003.0013.

Cameron AG, Birth of a Solar System, *Nature*418(2002)924.
Carlin JL, Mutations are the raw materials of evolution, *Nature Education Knowledge*3(2011)10.

Chandrashekar CR, *Mind your mind*, Nava Karnataka Publications Pvt. Ltd., Bengaluru, India (1987).

Chen Y, Lyga J, Brain-skin connection: stress, inflammation and skin aging, *Inflammation and Allergy Drug Targets*13 (2014) 177.

Celho P, *The Alchemist*, Thorsons, Harper Collins Publishers, London (2018).
Daly RT, Schultz PH, The delivery of water by impacts from planetary accretion to present, *Science Advances*4 (2018) eaar2632;DOI:10.1126/sciadv.aar2632.
Darwin C, *The Origin of Species*, Peacock Books, New Delhi (1859).
Davis A, *Medicines by Design*, National Institute of Health Publication No. 06−474 (2006).

Eigen M, Winkler-Oswatitsch R, *Steps Towards Life: A Perspective on Evolution*, Oxford University Press, UK (1996).

Feldman R, Magori-Cohen R, Galili G, Singer M, Louzoun Y, Mother and infant coordinate heart rhythms through episodes of interaction synchrony, *Infant Behaviour and Development* 34 (2011) 569.

Ferreri L, Mas-Herrero E, Zatorre RJ, Ripollés P, Gomez-Andres A, Alicart H, Olivé G, Marco-Pallarés J, Antonijoan RM, Valle M, Riba J, Rodriguez-Fornells A, Dopamine modulates the reward experiences elicited by music, *Proceedings of the National Academy of Sciences, USA* 116 (2019) 3793.

Frankel OH, Variation, the essence of life, *Proceedings of the Linnean Society of New South Wales* 95 (1970) 158.

Frankham R, Genetics and extinction, *Biological Conservation* 126 (2005) 131.

Frantz C, Stewart KM, Weaver VM, The extracellular matrix at a glance, *J. Cell Sci.* 123 (2010) 4195–4200; DOI: 10.1242/jcs.023820.

Ghosh G, Panicker L, *Journal of Nanoparticle Research* 16 (2014) 2800.

Ghosh G, Targeted ion therapy using nanoparticles–a new direction, *Atlas of Science*, February 1 (2016).

Ghosh G, Panicker L, Naveen Kumar N, Mallick V, Surface plasmon resonance of counterions coated charged silver nanoparticles and application in bio-interaction, *Materials Research Express* 5 (2018) 055005.

Ghosh G, Application of functional metal nanoparticles for biomarker detection, Chapter 3 in Biosensors: Materials and Applications, Ed. Inamussin, Rangreez TA, Ahamed MI, Asiri AM, *Materials Research Foundations* 47 (2019) 77−130, Materials Research Forum LLC, Millersville, PA 17551, USA.

Gilbert W, Origin of life: The RNA world, *Nature* 319 (1986) 618.
Goswami A, *Quantum Doctor: A physicist's guide to health and healing*, Jaico Publishing House, India (2018).

Grosberg RK, Strathmann RR, The evolution of multicellularity: a Minor major transition? *Annual Review of Ecology, Evolution, and Systematic s* 38 (2007) 621.

Gulkowski S, Olchowik JM, Hydrogen interactions in magnetic resonance imaging: Histogram-based segmentation of brain tissues, *Materials Science, Poland* 24 (2006) 1051.

Hatton L, Warr G, Protein structure and evolution: Are they constrained globally by a principle derived from Information Theory? *PLOS ONE* (2015) 0125663; DOI:10.1371/journal.pone.0125663.

Hawking SW and Mlodinow L, *The Grand Design*, Bantam Press, Great Britain (2010).

Inamuddin, Rangreez TA, Ahamed MI, Asiri AM, Biosensors: Materials and Applications, *Materials Research Foundations*, <u>47</u> (2019) Materials Research Forum LLC, Millersville, PA 17551, USA.

Keen JS, Consciousness, the laws of physics, the big bang, and the structure of the universe, *NeuroQuantology: An Interdisciplinary Journal of Neuroscience and Quantum Physics*<u>11</u> (2013) 227; DOI: 10.14704/nq.2013.11.2.610.

King N, The unicellular ancestry of animal development. *Developmental Cell* <u>7</u> (2004) 313.

Krauss LM, *A Universe from Nothing: Why There Is Something Rather Than Nothing*, Simon & Schuster, Great Britain(2012).

Krebs J, Hillebrandt W, The interaction of supernova shock fronts and nearby interstellar clouds, *Astronomy and Astrophysics* 128(1983) 411.

Kumar AM, Tims F, Cruess DG, Mintzer MJ, Ironson G, Loewenstein D, Cattan R, Fernandez JB, Eisdorfer C, Kumar M, Music therapy increases serum melatonin levels in patients with Alzheimer's disease, *Alternative Therapies in Health Medicine*<u>5</u>(1999)49.

Kwok S, *Physics and chemistry of the interstellar medium*, University Science Books, USA (2006).

Leca-Bouvier B, Blum LJ, Biosensors for protein detection: A review, *Analytical Letters* 38 (2005) 1491;https://doi.org/10.1081/AL-200065780.

Lennard-Jones JE, On the determination of molecular fields, *Proceedings of Royal Society of London* A 106(1924) 463.

Libby E, Conlin PL, Kerr B, Ratcliff WC, Stabilizing multicellularity through ratcheting, *Philosophical Transactions of the Royal Society of London, Series B: Biological Sciences* 371 (2016) 20150444.

Lindemann B, Receptors and transduction in taste, *Nature* 413 (2001) 219.

Lipska BK, *The neuroscientist who lost her mind: A memoir of madness and recovery*, Corbi Books edition, Transworld publishers, London, UK (2019).

Lipton BH, *The Biology of Belief: Unleashing the Power of Consciousness, Matter and Miracles*, Hay House Publishers (India) Pvt Ltd (2018).

Lipton BH, Bhaerman S, *Spontaneous Evolution: Out Positive Future and a Way to get There from Here*, Hay House Publishers (India) Pvt Ltd (2017).

Lunine JI, Physical conditions on the early Earth, *Philosophical Transactions of the Royal Society B: Biological Sciences* 361(2006) 1721.

Maeshima K, Matsuda T, Shindo Y, Imamura H, Tamura S, Imai R, Kawakami S, Nagashima R, Soga T, Noji H, Oka K, Nagai T, A transient rise in free Mg^{2+}ions released from ATP-Mg hydrolysis contributes to mitotic chromosome condensation, *Current Biology* 28 (2018) 444.

Martin WF, Garg S, Zimorski V, Endosymbiotic theories for eukaryote origin, *Philosophical Transactions of the Royal Society of London Series B: Biological Sciences* 370 (2015) 20140330. Misteli T, Where the nucleus comes from, *Trends in Cell Biology* 11 (2001) 149.

Montague PR, Neuroeconomics: a view from neuroscience, *Functional Neurology* 22 (2007) 219.

Mukherjee S, *The Laws of Medicine*, Simon & Schuster UK Ltd., London (2015).

Müller U, Cadherins and mechanotransduction by hair cells, *Current Opinion in Cell Biology* 20 (2008) 557.

Nickels M, Xie J, Cobb J, Gore JC, Pham W., Functionalization of iron oxide nanoparticles with a versatile epoxy amine linker, *Journal of Materials Chemistry* 20 (2010) 4776.

Nusbaum EC, Silvia PJ, Shivers and timbres personality and the experience of chills from music, *Social Psychology and Personality Science* 2 (2010) 199; https://doi.org/10.1177/1948550610386810.

Oosting PH, Signal transmission in the nervous system, *Reports on Progress in Physics* 42 (1979) 1479.

Papadimitriou A, Priftis KN, Regulation of the hypothalamic-pituitary-adrenal axis, *Neuroimmunomodulation* 16 (2009) 265.

Parkinson C, Kleinbaum AM, Wheatley T, Similar neural responses predict friendship, *Nature communications* 9 (2018) 332; DOI: 10.1038/s41467-017-02722-7.

Perry BD, Maltreatment and the developing child: How early childhood experience shapes child and culture, *The inaugural Margaret McCain lecture*, September 23 (2004).

Rokas A, The origins of multicellularity and the early history of the genetic toolkit for animal development, *Annual Review of Genetics* 42 (2008) 235.

Salimpoor VN, Benovoy M, Larcher K, Dagher A, Zatorre RJ, Anatomically distinct dopamine release during anticipation and experience of peak emotion to music, *Nature Neuroscience* 14 (2011) 257.

Schmandt B, Jacobsen SD, Becker TW, Liu Z, Dueker KG, Dehydration melting at the top of the lower mantle, *Science* 344 (2014) 1265;DOI:10.1126/science.1253358.

Siegel M, Engel AK, Donner TH, Cortical network dynamics of perceptual decision-making in the human brain, *Frontier of Human Neuroscience* 5 (2011) 1.

Siemens PJ, Jensen AS, *Elements of Nuclei: Many-Body Physics with the Strong Interaction*, CRC Press, Taylor & Francis Group, 6000 Broken Sound Parkway NW, Suite 300, Boca Raton, FL, USA(2018).

Sparke LS, Gallagher III JS, *Galaxies in the Universe: An Introduction*, Cambridge University Press, UK(2000); ISBN 978-0-521-59740-1.

Stearns SC, Ackermann M, Doebeli M, Kaiser M, Experimental evolution of aging, growth, and reproduction in fruit flies, *Proceedings of the National Academy of Sciences, USA* $\underline{97}$(2000) 3309.

Sunderland M, *The Science of Parenting*, DK Publishing, London, UK 2006).

Voet JG, Voet JD, *Biochemistry* (3rd ed.), New York: John Wiley & Sons(2004); ISBN 978-0-471-19350-0.

Ward P, Kirschvink J, *A New History of Life: The Radical Discoveries about the Origins and Evolution of Life on Earth*, Bloomsbury Pres s(2015); ISBN 978-1608199105.

Wilczewska AZ, Niemirowicz K, Markiewicz KH, Car H, Nanoparticles as drug delivery systems, *Pharmacological Reports* $\underline{64}$ (2012) 1020.

WEB RESOURCES:

https://www.ck12.org/biology/Significance-of-Carbon/lesson/Significance-of-Carbon-BIO/

https://www.singerinstruments.com/resource/what-are-genetic-mutatio/

https://onekindplanet.org/top-10/top-10-worlds-extinct-animals/

https://www.brainpickings.org/2012/04/27/when-einstein-met-tagore/

https://www.bellybelly.com.au/baby/mothers-and-babies-synchronise-brainwaves-when-making-eye-contact/

https://www.sharecare.com/health/music-therapy/can-music-heal-disease -mental/

https://bebrainfit.com/music-brain/

http://hyperphysics.phy-astr.gsu.edu/hbase/Particles/quark.html

NOTES

[1]**Light year** is the distance traveled by light at a speed of 300,000 m/s in 1 year $\approx 9.5 \times 10^{12}$ mile.

[2]When an electron enters into the attractive field of a proton; it has to give away its positive energy (of free existence) which is emitted as a photon.

[3]Proton and neutron can interchange their identities, i.e., proton can become neutron and neutron can become proton, through changing the states of their constituent quarks.

[4]**Bosons**, named after the pioneering work of Satyendra Nath Bose, are particles that follow Bose-Einstein statistics. Gauge bosons are boson particles whose interactions follow gauge theory.

[5]A periodic arrangement of atoms, molecules or ions is called **lattice**.

[6]**Aromaticity** is a property of a cyclic, planar organic molecule with a ring of resonance bonds that gives it better stability compared to other geometric or connective arrangements with the

same set of atoms.

[7]**Stem cells** are used in therapy to repair response of diseased, dysfunctional or injured tissues. It is a modern therapeutic technique and is also known as regenerative medicine.

[8]**Lymph nodes** are small lumps of tissue that contain white blood cells, which fight infection. They filter harmful substances like bacteria and cancer cells from our body, and help in fighting infections. They also play an important role in cancer diagnosis, treatment and prognosis.

[9]The **Copenhagen interpretation** is an interpretation of quantum mechanics; it suggests that a quantum state, which is represented by a wave function, can have many probable states. It was developed by Werner Heisenberg, Niels Bohr and Albert Einstein in Copenhagen in 1927.